Aspects of Science and Technology in Ancient India

This book critically examines different aspects of scientific and technological development in Ancient India. It studies the special contribution of the history of science to our scientific understanding and its relationship with the philosophy and sociology of science.

The volume:

1. Discusses diverse and wide-ranging themes including the Tibetan Buddhist tradition of neuro-biology
2. Sheds light on the unique developments within iron technology and urbanization in ancient Odisha
3. Studies the trajectory of proto-historic astronomy in India and the science of monsoon in early India
4. Evaluates the legacy of Aryabhata based on his major works related to astronomy and mathematics through a multidimensional perspective
5. Analyzes the traditional knowledge of medicine in early India, the golden age of surgery with reference to the ancient Greek and Arabic systems of medicine, and the Buddhist influence on the science of medicine in Tibet

This book will be an essential read for scholars and researchers of ancient history, Indian history, history of science, history of technology, science and technology studies, and South Asian studies.

Arun Kumar Jha has had a long teaching and research experience as Professor of History in T. M. Bhagalpur University, Bhagalpur. He worked as Visiting Fellow and Visiting Professor, besides being Resource Person in various universities. He was empanelled by the Indian Council for Cultural Relations (ICCR), New Delhi, as Professor to the Chairs of Indian Studies Abroad. There are several published research papers and books to his credit.

Seema Sahay is an accomplished academic writer and a freelance consultant associated with the publishing industry. She holds an impressive academic record up to her Ph.D. and has worked as a Junior Research Fellow of Indian Council of Historical Research (ICHR), New Delhi, in the Department of History, Tilkamanjhi Bhagalpur University, Bhagalpur. She is engaged in historical studies especially related to the science of medicine in the history of Ancient India. She has participated in a number of seminars and conferences. Some of her research papers were published in different journals and proceedings.

Aspects of Science and Technology in Ancient India

Edited by
Arun Kumar Jha and Seema Sahay

Routledge
Taylor & Francis Group
LONDON AND NEW YORK

First published 2023
by Routledge
4 Park Square, Milton Park, Abingdon, Oxon OX14 4RN

and by Routledge
605 Third Avenue, New York, NY 10158

Routledge is an imprint of the Taylor & Francis Group, an informa business

British Library Cataloguing-in-Publication Data
A catalogue record for this book is available from the British Library

Library of Congress Cataloging-in-Publication Data
A catalog record has been requested for this book

ISBN: 978-1-032-24859-2 (hbk)
ISBN: 978-1-032-44138-2 (pbk)
ISBN: 978-1-003-37064-2 (ebk)

DOI: 10.4324/9781003370642

Typeset in Sabon
by Deanta Global Publishing Services, Chennai, India

In loving memory of my parents, Jayadeva Jha and Sita Devi,
who are always there with us.

Contents

Contributors

Deepak Kumar has taught history of science and education at the Zakir Hussain Centre for Educational Studies, Jawaharlal Nehru University, New Delhi. He also held a concurrent professorship at the Centre for Media Studies at the university. He is the author of *Science and the Raj* and has co-edited several books relating to history of education, medicine, technology and environment in colonial India. During almost four decades of research and teaching in India, the author has tried to popularize the History of Science, Technology, Environment & Medicine (HISTEM) in several universities in India and abroad as well.

Arun Bandopadhyay has a wide range of research interests: agrarian history, business history, social and cultural history, and history of science and environment, with contributions to each of them. He retired as Nurul Hasan Professor of History and Dean of Arts, University of Calcutta. Currently, he is the Historical and Archaeological Secretary of the Asiatic Society, Kolkata. His latest books include *Science and Society in India, 1750-2000* (New Delhi, 2010), *Nature, Knowledge and Development: Essays in Environmental History of India* (New Delhi, 2016), *Prabuddha Bharat: Understanding Ambedkar in the Passage of Time* (Kolkata, 2020), and two edited works (with Amiya Kumar Bagchi) on *Documents on Economic History of Northern India in the Late Nineteenth Century: Quality of Life*, Volume I: 1860s–1870s, in three Parts (Delhi, 2020), and Volume II: 1880s–1890s, in three Parts (Delhi, 2022).

Amrendra Kumar Thakur is Professor in history at North-Eastern Hill University in Shillong, Meghalaya. His research interests include the history of medieval India and pre-colonial Northeast India. He has published several reviews, essays and books, including *Technology of the Tribes of Northeast India* (2017). Till 2019, he served as the General Secretary of North East India History Association, one of India's premier organizations devoted to historical study and research.

Byomkesh Tripathy teaches History at Indira Gandhi National Tribal University, Amarkantak, Madhya Pradesh. He specializes in the field of ancient history, culture and archaeology. He has contributed to Buddhism and history of tribes of Northeast India. He has also authored a number of books and research papers to his credit. Presently, he holds the post of the Vice Chancellor of Utkal University of Culture, Bhubaneswar, Odisha.

Dibishada Brajsundar Garnayak, before joining Archaeological Survey of India as a Deputy Superintending Archaeologist in 2015, he served as Assistant Professor in Asmara University, Eritrea, and Associate Professor in Mekelle University, Ethiopia. Currently, he is posted as Superintending Archaeologist in newly created Meerut Circle in Uttar Pradesh. Dr. Garnayak is a promising Field Archaeologist who has published more than a dozen research papers in various national and international journals based on his field works in both the African countries he served. He has also authored monographs entitled *Early Historic Archaeology of Odisha*, *Rock Art in Coastal Odisha* and *Prehistoric Art in Odisha* and has also edited a book entitled *Rock Art in India*. His research interests are art and architecture, besides field archaeology.

Sabyasachi Chatterjee was the President of the All India People's Science Congress (AIPSN) from 2015 to 2022. He has been with the AIPSN since its inception, focusing mainly on science communication and self-reliance in science and technology. He is a retired scientist from the Indian Institute of Astrophysics (IIA), Bengaluru. From 1980 to 1982, he was a Visiting Scientist at the International Centre for Theoretical Physics, Trieste, Italy. His research areas have been in solid and liquid states, galactic structure and astrophysical and medical imaging.

Venkateswaran Thathamangalam Viswanathan, a science communicator, science writer and founder chief editor of *India Science Wire*, is a senior scientist, Category F, with "Vigyan Prasar", DST, National Institute for Science Communication, New Delhi. His research profile includes circulation of modern science in pre-modern Tamil region, the evolution of calendars and history of astronomy and science, colonialism and modernity in India, evolution of science content in school textbooks, history of science communication in Indian cultural cosmos, and mediation between science, society and media.

Amarnath Thakur, after obtaining Ph.D. in 1988 from the University of Delhi, started teaching at post-graduate level for almost a decade in the University of Delhi before switching to the Indian National Science Academy (INSA), New Delhi, to head its History of Science Division. Having authored several books and numerous research papers, he superannuated from services in 2018 and is currently engaged in several social and educational initiatives.

Seema Sahay is an accomplished academic writer and a freelance consultant associated with the publishing industry. She holds an impressive academic record up to her Ph.D. and worked as Junior Research Fellow of Indian Council of Historical Research (ICHR), New Delhi, in the Department of History, Tilkamanjhi Bhagalpur University, Bhagalpur. She is engaged in historical studies especially related to science of medicine in the history of Ancient India. She has participated in a number of seminars and conferences. Some of her research papers were published in different journals and proceedings.

Melarkode Narayanaswamy Rajesh, after completing his post-graduation and M.Phil. from the Centre for Historical Studies, Jawaharlal Nehru University (JNU), New Delhi, he did his Ph.D. on the Buddhist monasteries of Tibet (Gompas) at the Centre for Central Asian Studies, School of International Studies, JNU. He has also worked on the socio-religious context of the early Chola Temples. He is presently working as Assistant Professor in History at the Department of History, University of Hyderabad. His main areas of research include the history of Tibet and religious movements in medieval South India and international affairs related to Central Asia.

Arun Kumar Jha has had a long teaching and research experience as Professor of History in T. M. Bhagalpur University, Bhagalpur. He worked as Visiting Fellow and Visiting Professor, besides being Resource Person in various universities. He was empanelled by the Indian Council for Cultural Relations (ICCR), New Delhi, as Professor to the Chairs of Indian Studies Abroad. There are several published research papers and books to his credit.

Acknowledgements

I am grateful to the Indian Council of Historical Research, New Delhi, for providing financial assistance, and to the Vice Chancellor of T. M. Bhagalpur University, Bhagalpur, for the fund from the University Grants Commission (unassigned) grants that enabled us to conduct the seminar on "Science and Technology in Ancient India (c. 2500 BC–1200 AD)". We would like to thank all the contributors who held their patience throughout the entire process of publication of this volume. We would like to convey our sincere gratitude to Prof. Deepak Kumar, Professor of History of Science and Education, Zakir Hussain Centre for Education (ZHCE), JNU, New Delhi, for his special efforts and encouragement without which the publication of this volume would not have seen the light of the day. Prof. Ishrat Alam, Former Member-Secretary, ICHR, New Delhi, and Professor and Head, Centre for Advanced Studies (CAS), Department of History, Aligarh Muslim University, Aligarh, also deserves our deep thanks for his interest in promoting the academic seminar and for encouraging us to take up this publication. Besides, credit goes to Seema Sahay, Associate Editor, for ensuring the publication of this work.

I also appreciate the insightful discussions and suggestions of my friends and family, especially Dr. Hem Shankar Sharma, an eminent and esteemed Physician and Professor of Medicine, J. L. N. Medical College and Hospitals, Bhagalpur (Bihar); Shri S. K. Tyagi, Ex-Curator, Department of Ancient Indian History, Culture & Archaeology, Tilkamanjhi Bhagalpur University, Bhagalpur; Shri Binoy Kant Pathak, Ex-A.Z.M, LIC of India, Bhopal Division, MP, and formerly a Professor of Mathematics, S. P. College, Dumka (Jharkhand); Dr. Arun Kumar Jha, the celebrated Surgeon and Ex-Deputy Director, Health Services Bihar, Patna, and last but not the least, Rashmi Mishra, an MBA, and Nishith Vatsa, Sales and Business Development Manager, Union Rebar Company, Abu Dhabi, UAE, who also went tirelessly through all the pages of the manuscripts. Finally, I feel pleased to thank my spouse, Sudha Jha, and my granddaughter, Shanaya N. Vatsa, who always support my passions and dreams.

Introduction

The journey within takes you to the infinite possibility of the Cosmos.[1]

> "It has recently been argued that a revolution in scientific knowledge comes about not through the accumulation of data alone but through a change in the paradigm. When the framework of explanation ... or a new set of questions are posed only then can there be a breakthrough in scientific knowledge."[2] This applies as much to the history of science and technology in ancient India. The assertion that the Indian philosophy is simply based on spiritualism and mysticism to the negation of mundane affairs emanates from the colonial Orientalist perception of the Indian past which "obscures a rich tradition of scientific and humanistic thought and analysis in early India."[3] This perception of colonial historiography had a definite political objective and extreme eurocentrism which the Westerners believed that the Orients could never emerge out of the esoteric cocoon of metaphysics and mysticism.[4] But the Western pronouncements can in no way arbitrate the nuances of our cultural inheritance. We are well aware that centuries ago Said-al-Audalusi in his *Tabaqat-al-Uman* has depicted India as a treasure trove of many of the scientific ideas which were cultivated first in ancient India.[5]

This volume comprises seminar presentations related to the tradition of techno-scientific outlook which gradually evolved as a result of the politico-cultural interactions of ancient Indians with the outside world as well as social changes within the region.[6] This techno-scientific approach may be defined "as technical artifact and science as knowledge. Both are historical variables."[7] Science in part is knowledge and so is technology, and both can be intertwined to form the cognition base of the techno-scientific paradigm.

For long, ancient India has been reckoned indistinguishable in the ranking of eminence in the context of the development of science and technology in the history of mankind. Fortunately, a host of distinguished scholars and historians, both local and foreign, such as K. R. L. Gupta, Girindranath Mukhopadhayaya, Prafulla Chandra Ray, B. N. Seal, D. M. Bose, S. N. Sen, B. V. Subbarayappa, Irfan Habib, Debiprasad Chattopadhyaya, Abdur Rahman,

DOI: 10.4324/9781003370642-1

D. P. Agrawal, Deepak Kumar, Dhruv Raina, A. K. Bag, H. C. Bhardwaj, Vibha Tripathy, and outside India, some of the prominent luminaries who did pioneer works are Henry Thomas Colebrook, George Thibaut, Julius Jolly, Rev. Ebenezer Burgess, Heinrich Robert Zimmer, Takao Hayashi, etc. – the significant contributions of all these social scientists led to wider perceptions of the history of science in India which, later on, became a major discipline.

A natural question that arises here is, what does a historian have to do with science and technology? Apart from a simple narration of the course of events in the development of the subject, what sort of contribution may a historian be expected to make by way of objective and analytical interpretation of the fundamental rudiments and propositions of science? One has to understand that this discipline does also have a social bearing. Any discovery, invariably in science and technology, has had always a socio-historical context. In fact, the invention in agricultural technique, animal husbandry and the machine-driven Industrial Revolution are logically interpreted both as cause and effect of socio-historical compulsions. Historiography of science and technology also has an applied facet and dimension. It enables us to understand its role in transforming the societies of yore and, on that basis, equips us with a better technique to plan our future course of action with regard not only to striking a balance between nature and sustainable evolution but also to emphasizing its social role through history. As Albert Einstein rightly remarks, "The whole of science is nothing more than the refinement of everyday thinking."[8] This is true to everyday life and living also.

In fact, the history of civilization in its material aspect is largely based on the scientific progress which remoulds the entire archetype of human thought. It is in this background that the historical study of science is essential to comprehend a man in relation to his society and physical environment. A close perusal of the successive stages of socio-cultural development from the hunting–gathering mode of human life to food production, urbanization and, thereafter, the rise of human civilizations or empires and nation-states is decidedly shaped by the progress made in science and technology. Any study without taking them into consideration would be partial and inadequate. So, it is imperative to have an objective evaluation of the historical studies on various aspects of science and technology.

When we go through the texts,[9] both religious and secular, and when we study the relics of India through the ages, we come to know of its material cultures and its concerns for the quality of life.[10] Like the architectural marvels of the past and specimens of arts and literary works, the advancement made in the arena of science and technology, such as the *Sulvasutras*, the Sudarshan lake of the Mauryan period, the Kushana water filtration tank, Chandra's rustless Iron Pillar, Jeevaka, Charaka, Sushruta, Aryabhata-I, Varahmihira, Brahmagupta, Bhaskaracharya-II, etc. are symbols of our scientific heritage.

It would be quite pertinent and germane to the purpose to have a glimpse of the history of science in ancient times.

The earliest indigenous cultures which interest the historians of science are those centred upon the mature phase of the Harappans (c. 2500 BC–1750 BC). The planned layouts of the Harappan cities show that civil engineering and town planning were quite advanced. The system of irrigation, river dams, water storage tanks, etc. was remarkable. Moreover, a developed waterproofing technology as shown by the Great Bath at Mohenjodaro and acquaintance of the knowledge of hydraulic engineering at Dholavira by using the Ground Penetration Radar technique are some of the striking features of the Harappans.[11] The monumental remains of fortifications, granaries, sewage systems, roads and house blocks excavated at Harappa, Mohenjodaro, Kalibanga and Lothal imply a good deal of arithmetic and knowledge of the geometry of rectangles, parallels and circles.[12] Besides, some of the artefacts like seals displaying intricate geometric patterns, a designed ruler calibrated to ten equal parts, bricks having the ratio 1:2:4 regardless of their size and location, etc. suggest, in turn, a methodical system in weights and measures which have been collected at several sites.[13] In fact, this system of weights is unique in the ancient world and there is much to learn from the Harappans about water management in arid areas. We further note the development in the sphere of textiles, pottery and metals and alloys.[14]

With the onset of the Vedic age (c. 1500 BC–c. 500 BC), the growth of Indian science is influenced by the speculative and philosophical mind. The contemporary literature, *viz.* the Vedas, Vedanga, Vedanta, Upanishads and Brahmanas, serves as the only source for the objective assessment of scientific development of the period. Rudimentary concepts of astronomy and mathematics can be found in these Vedic writings. But a substantial knowledge of measurement and geometry appears in the *Sulvasutras*[15], which are a collection of texts written during 1000 BC–500 BC.[16] These works discuss the construction of fire altars of different geometrical shapes of the same as defined for each god. They also illustrate the methods of geometric area-preserving transformations from one geometric shape to another. We also find here a method of the inverse transformation.[17] Furthermore, the Baudhayana and Apastamba sutras give a value for $\sqrt{2} = 1.4142156$ while the correct value for $\sqrt{2}$ is 1.414212. The ancient and modern values are almost the same. G. Thibaut is right in his observation that "thus it is clear that the ancient Hindus have attained a remarkable degree of accuracy in calculation of the approximate value for $\sqrt{2}$."[18] Many foreign mathematicians have written commentaries on the sutras, e.g. G. Thibaut, M. Winternitz, C. Muller, etc.

The concept of cosmology and astronomy (*Jyotisha/Vedanga*) was in a formative stage during the period of the Vedas. It was used for the purpose of agriculture and animal husbandry and to settle the date and time for performing the periodical sacrifices. The lunar mansion along the path of the sun and the moon were known to the Vedic people. A year comprising 12 solar months and 6 seasons was recognized by them.[19]

The nature of the healing tradition in the Vedas is mostly psychosomatic. The earliest mention of medicines is in the *Atharvaveda*. Here the remedies recommended by *bhishak* (physician) are magical charms and spells. Besides, this treatise refers also to the use of certain plant-derived products to treat different diseases, i.e. fever, cough and cold, indigestion, headache, tumour.[20]

For long, it was thought that the early Indians were not interested in the materialistic line of thinking. This does not hold true in the context of the *Samkhya* system of Kapila, the metaphysics of *Nyaya-Vaisheshika Sutras* of Gautama-Kanada, the materialism of Charavaka's *Lokayata* and the rise of Jainism and Buddhism which denied the authorities of the Vedas. The presence of such a rich array of philosophical currents and clash of ideas created an intellectual fervour which precipitated a resurgence of science beginning in the 6th century BC. These factors led to an atmosphere of scientific enquiries with reference to different areas of human endeavours.

Some of the great illuminating thinkers of the time namely Aryabhata-I (5th century AD), Varahmihira (6th century AD), Brahmagupta (AD 598–668), Bhaskara-I (a contemporary of Brahmagupta), Mahavira (c. AD 850), Aryabhata-II (c. AD 950), Bhaskara-II (AD 1114–1185) really enriched the realm of knowledge. In the field of mathematics, there appeared three distinct contributions of the ancient Indians: the notation system, the decimal system and the use of zero.[21] And in the arena of astronomical practices, the calculation of the planetary system, the rotation of the earth on its axis and revolving around the sun, the precession of the equinoxes and the true causes of eclipses – all these facts were known.[22]

Regarding the earliest use of decimal notation in literature, P. V. Kane says,

> In view of the discovery of decimal place value concepts in India, it is accepted that 0 (zero), as a part of the numerical system, is an Indian contribution. Pingalacharya's *Chanda Sastra* first mentioned the word *Soonya*, and also described the rules for calculating the number of long and short syllables in metre of different syllables.[23]

The *Chanda Sastra* has been recorded as the most ancient Sanskrit treatise on prosody. It has been dated either to the later part before the Christian Era or to the beginning of the Christian Era. This was the time when an evident transition took place from the Vedic metre to the classical metre in Sanskrit literature.

However, the earliest epigraphic evidence of the place value system in Indian inscriptions is found in the Mankani plates from Gujarat dated AD 595–596. The next reading occurs in the Gwalior inscription dated AD 791–792. Zero appears as a dot in an 8th century plate and as a full-drawn circle in plates of AD 870 and AD 876.[24] There is no evidence of the use of the above system of numerals including zero in a decimal place value scale as it was found in India. By AD 718, the Chinese learnt this Indian notational system from the Buddhist missionaries. In Arabia, by c. AD 825, Al-Khwarizmi wrote a treatise named *Tafriq bin-Hisab-al-Hindi* (or in English, it is known as '*On the Calculation within Hindu Numerals*'). The Arabic word '*hisab*' (*al-hasb*) means to count or number and refers to calculation or arithmetic.[25] Here, it is interesting to learn that the word '*hisab*' gradually grew in popularity and became synonymous with '*ganita*' (mathematics) in India since medieval times. In AD 830, Al-Kindi wrote four volumes *On the Use of the Indian Numerals*.[26] These works were largely responsible for the diffusion of the Indian system of numeration in the Middle East and the West. According to George Alfred Leon Sarton, "our numerals and the use of zero were invented by the Hindus and transmitted to us through the Arabs and hence known as Arabic numerals which we have given them the name."[27] This flow of knowledge, from India to the outside world, nurtures the scientific cultures of today.

While the urge behind the mathematical research came mainly from problems of ordinary life, Aryabhata-I (AD 499)[28] extended his mathematical method to solve problems of astronomy which he presented in an epicyclical model.[29] His *Aryabhatiya* describes several ideas that are foundational to modern astronomy and mathematics. Its *Ganitapada* section covers mensuration, arithmetic and geometric progressions, simple algebra, etc. Its *Kalakriyapada* enumerates different units of time, and *Golapada* puts in geometric and simple trigonometric aspects of the celestial sphere, etc.[30] Aryabhata worked on the approximation for pi (π) in the *Ganitapada* and holds that $\pi = \frac{62832}{20000} = 3.1416$, which is accurate to four decimal places. His value of π is a very close approximation to the modern value and the most accurate among those of the ancients.[31] After *Aryabhatiya* was translated into Arabic (c. AD 820), this approximation was mentioned in Al-Khawrizmi's book also.[32] It is further interesting to note that Aryabhata's methods of astronomical calculations have been in continuous use for practical purposes of fixing the *Panchangam* (Hindu Calendar). The extreme brevity of the text was elaborated in commentaries by his disciples Bhaskar-I, in his *Bhashya*, and by Nilakantha Somayaji, in his *Aryabhatiya Bhashya* (AD 1465).

Varahmihira (c. AD 550) also maintained the traditions of Aryabhata-I in his writings entitled the *Panchasiddhantika* and the *Brihatsamhita*. He summarized his ideas related to mathematics and astronomy in his *Panchasiddhantika*, the five Siddhantas named Paitamaha, Vasishtha, Paulisa, Romaka and Surya. Some scholars refer to *Panchasiddhantika* as the old *Suryasiddhantika* and date it to AD 505.[33] Highly regarded and widely disseminated, the *Suryasiddhanta* had a profound influence on the course of astronomy especially rules to calculate the motions of various planets, the moon and the orbits of various astronomical bodies.[34]

Regarding the tradition of Indian decimal notation, it reached its high watermark with Brahmagupta and his *Brahmasphuta-Siddhanta* (AD 628), where its 12th and 18th chapters refer to important mathematical developments, i.e. cyclic quadrilaterals, square and cube roots, rule of three, progressions. In his later work, *Khandakhadyaka* (AD 665), he explains the method of simplifying and improving astronomical calculations.[35]

Some of the illustrious scholars such as Bhaskaracharya-I (7th century AD), Lallacharya (AD 748), Manjulacharya (AD 932), Aryabhata-II (AD 950) and Sripati (AD 1039) also enhanced the faculty of human knowledge by contributing to different disciplines of natural sciences. One of the prominent philosophers in c. AD 850 was Mahaviracharya who wrote *Ganitasarasamgraha* which is the first textbook on arithmetic in the present day. It describes in detail the current method of finding the Lowest Common Multiple.[36] Another mathematician, Shridhara (c. 8th century), takes up a number of complex problems relating to arithmetic, mensuration and geometry in his *Patiganita.*[37] The earliest example of this kind of presentation including algebraic signs, etc. may be noticed in the Bakhshali manuscript (c. 7th century AD). It is an incomplete ancient Indian mathematical work written in old Sarda characters on 70 leaves of birch bark that was found in 1881 in the village of Bakhshali near Peshawar. It is currently housed in the Bodleian Library of the University of Oxford.[38]

The method of indeterminate analysis reached its zenith in the works of Bhaskaracharya-II, i.e. *Lilavati* and *Bijagantia*, which form part of the *Siddhanta Siromani* (AD 1150). He described the solution of the first-degree equation in terms of the pulverizer (*Kuttaka*).[39] By introducing the solution of the general equation through *chakravala* or cyclic method, he has perpetuated his name for all the time to come in the history of the theory of numbers.[40]

Much of Indian mathematics and astronomy reached the Arabic world between the 8th and 11th centuries and thence travelled to medieval Europe.[41] In the 8th century of Arabia, the book *Aryabhatiya* was known as *Arjbar-Sindhind*, and the classical astronomical and mathematical traditions were critically maintained in Abu Raihan al-beruni's *Kitab-al-Hind* (c. AD 1035). A Persian version of *Lilavati* appeared in 1587 on the order of Akbar and one of *Bijaganita* in 1635 for Shah Jahan.[42]

In ancient India, the healing tradition gradually grew from folklore to a sequence of science. The management of relief from pains became the motto which promoted the idea of acquiring knowledge of medicine and its application. By the c. 4th century AD, India produced two distinguished personages of Ayurveda (the science of longevity) – Charaka, the physician, and Sushruta, the surgeon. Their writings, Samhitas, suggest a fully evolved medical tradition whose holistic healing approach may approximately be analogized to those of the Hippocratic and Galenic systems. In fact, the practices of yoga and the mystical experience of the ancient healers along with Buddhism influenced the study of physiology which enriched their learning.[43] The *Samhita* of Charaka is a repository of medical knowledge. It explains the symptoms of different types of fever, leprosy, hysteria and tuberculosis and also takes into account their remedies and cures as well. It enumerates the names of a large number of plants and herbs to be used as therapies. We also come to know of several ancient Indian flora and fauna.[44]

In AD 1890, we come across an old extant manuscript pertaining to medicine which is relevant to discuss here. It is discovered by a British Lieutenant, Hamilton Bower, in a Buddhist monument at Kucha near Xinjing in northwestern China.[45] It is dated between AD 500 and AD 550. It is written on birch bark which includes fragments of three medical treatises, and these have much in common with other Sanskrit texts of Charaka, Sushruta, Ravigupta, Vagbhata and Kashyapa. It is preserved in the collections of the Bodleian Library in Oxford.[46] It contains medical prescription sections on garlic, powder, medicated clarified butter (*ghee*), oil, elixirs, aphrodisiacs, decoctions, ointments, etc.[47] The discovery of the Bower manuscript in remote China near the Central Asian region is considered evidence of the spread and sharing of ideas in ancient times between India, China and Central Asia.[48]

Sushruta Samhita is considered to be one of the most comprehensive textbooks on ancient surgery in line with empirical studies.

> The caesarian section was known, bone-setting reached a high degree of skill, and plastic surgery was developed ... the repair of noses, ears and lips, lost or injured in battle or by judicial mutilation ... The pharmacopoeia ... was very large, and comprised animal, vegetable and mineral products.[49]

In this *Samhita*, we find the mention of specialized surgical equipment in common use which consisted of 20 types of knives and needles (*sastra*), 30 probes (*salaka*), 20 tubular instruments and 26 articles of dressing (*upayantra*), etc.[50]

The subsequent period also saw medical works such as Vagbhata's *Ashtangasamgraha*, *Ashtangahridayasamhita* of another Vagbhata, Madhavakara's *Rugvinishchaya* (concerning diagnosis and pathology) and Chakrapanidatta's *Chikitsasara-samgraha* (related to therapeutics).

The basis of Ayurveda is to keep the three primary bodily humors, *vata* (wind), *pitta* (bile) and *kapha* (phlegm), in balance. One of its important features is its insistence on medication through massage, meditation, yoga and dietary changes and the use of herbal and mineral remedies besides abstention if required. It also developed a fairly elaborate code of ethics for physicians.[51]

Charaka and Sushruta describe the properties of six metals: iron, tin, lead, copper, gold and silver. The medical chemists formulated in producing many alkalis, i.e. K_2CO_3, Na_2CO_3, acids and metallic salts by calcination and distillation. Books written in the later periods refer to the use of mercury, zinc and sulphur also.[52]

Nagarjuna is a name synonymous with Alchemy, Chemistry and Metallurgy. His *Rasaratnakara* and *Rasaratnasammuchya* are his renowned contributions to throwing light on Alchemy and related subjects. A detailed account of this knowledge can be found in Acharya Prafulla Chandra Ray's famous book '*The History of Hindu Chemistry*'.

We know from the evidence of the Iron Pillar of Delhi and other sources that metallurgy has a very rich tradition in ancient India. The extractions of metals from ores and the mixing of different metals to obtain alloys, i.e. brass, bronze and steel, were known and their products like Wootz steel, swords, weapons, etc. were in great demand in the Roman Empire and the Middle East.[53]

Regarding the healing system of Tibet, it is closest to the Ayurvedic tradition of India both in principle and in practice. It is based upon Indian Buddhist Literature (for example, *Abhidharmakosabhasyam* and *Vajrayanatantras* containing practices based on medical anatomy) and Ayurveda as well.[54] It embraces the traditional Buddhist belief that all illness results from the three poisons: delusion, greed and aversion. It follows the Buddha's Four Noble Truths which apply medical diagnostic logic to suffering.[55] Its *Four Tantras* is the primary work for Tibetan medical practice.[56] It employs a complex approach to diagnoses such as an analysis of pulse and urine, behaviour and dietary notifications, medicines composed of herbs and minerals and physical therapies to treat illness.

The present volume incorporates ten articles contributed by faculty of various streams like history, astronomy, mathematics and medicine in order to exchange their experiences and observations for the benefit of knowledge as well as society.

Deepak Kumar has his viewpoints concerning the distant past which always fascinates and often tends to border on romanticism. How does one place scientific and technological developments in our civilizational growth? What constitutes science in the context of our ancient past? Every society, however crude it might appear from the benefit of hindsight, has a certain amount of scientific rationality and endeavour within it. Science is intrinsic to human nature. In recent years, environmental issues have attracted the attention of everyone. It is no doubt an extremely important area of research requiring inter-disciplinary knowledge involving geography, geology, chemistry, zoology, medicine, anthropology and history.

Arun Bandopadhyay in his paper has discussed the different approaches to the study of history of science which underwent critical changes in the last two centuries before they could more definitively tilt towards the social dimensions of the making of science. Several methodological questions on the study of History of Science, in general, may be identified. This paper focuses on six such questions in connection with the relationship between the history of science and the study of science, the special contribution of the history of science in our scientific understanding, its relationship with the philosophy and sociology of science, its didactic role in demonstrating the true nature of scientific knowledge, its reflection on the humanist placement of science and its general importance as a field of study.

Amrendra Kumar Thakur highlights the topic of 'Re-reading of Iron in Northeast India' and presents his outlook with context to nationalists and communal approaches. He posits his ideas that unlike in India, the enquiries into the past of the people in Europe and America started about more than a century and a half back. Archaeologists, anthropologists, cultural historians and many social scientists, however, have been studying the technological progress of the past from their own perspectives; most of the time these were unidimensional. Nonetheless, all of them have contributed to the importance of technology in making sense of the past. This paper is based on archival, library and field researches and discusses various aspects of iron technology and its socio-cultural significance in Northeast India prevailing during the pre-colonial period. The paper, sometimes, goes beyond AD 1200 due to the features of the area of study.

Byomkesh Tripathy and Dibishada Brajsundar Garnayak shed light on a unique distribution pattern concerning iron technology and urbanization in ancient Odisha. They also point out that the knowledge of iron metallurgy by the inhabitants brought changes in the origin and growth of urban centres in the state. Besides, the landscape potentiality of the region and the richness of mineral resources led to the development of trade and commerce in the early historic period which ultimately paved the way for the first urbanization in Odisha. Hence, the authors mark the period from 200 BC to AD 100 as the turning point in the history of material culture and the prosperity of the people of the land.

Sabyasachi Chatterjee has evaluated the legacy of Aryabhata on the basis of his major works related to astronomy and mathematics. At the beginning of his article, he alerts us against those ideas which believe that modern science can be rediscovered from 'Hindu' spiritual books of antiquity. He strongly concurs with the concept of making an objective assessment of Aryabhata. Hence, the achievements of Aryabhata lie in the areas of astronomy, questions of time reckoning and spherical geometry and trigonometry. Moreover, the author maintains that it was the Chinese mathematician, Zhu Chongi (5th century AD), whose discovery of the value of *pi* was much more accurate than that of Aryabhata. The legacy of Aryabhata has been appreciated in the paper but, on the other, has been criticized as well for the inaccuracy of many of his predictions. Concludingly, the author accepts that Aryabhata's contributions to knowledge need to be evaluated from a multidimensional perspective.

Venkateswaran Thathamangalam Viswanathan has reviewed the origin of ancient Indian astronomy from a wider perspective. Different tools and methods of archaeo-astronomy and ethno-astronomy have been enumerated in the article to reveal the basics of the astronomical traditions dating back to the proto-historic period. The author sheds sufficient light on a number of subjects in this regard, i.e. the gradual transition to Neolithic Culture, scientist's perception of cosmogony, explanation of the sidereal year, astronomical import of orientation, heliacal rising of Pleiades, characteristics and astronomical significance of megaliths in the Mesolithic period. Besides, it is stated here that some or all of the megaliths have astronomical implications which are the first creative scientific pursuit. It has been further elucidated that megaliths are inceptive 'astronomical observatories' of the ancient world. But the author hastens to caution the readers to the notion of 'observatory' which is not the same as the modern meaning.

Amarnath Thakur has scientifically studied about Sushruta and the golden age of surgery in ancient India which rests largely on the accomplishments of Sage-surgeon, Sushruta. He practised and taught the art of surgery at the University of Banaras in the ancient city of the same name, located on the banks of the river Ganges. His monumental treatise on surgery, *Sushruta Samhita*, established him as the 'Father of Indian Surgery'. He was the first surgeon to systematize surgery by dividing it into separate fields. He is known as the originator of a cataract operation, laparotomy and vesical lithotomy. The paper further enquires about the connection of ancient Indian medicine, especially the surgical system with the contemporary Greek and Arabic systems of medicine which were in vogue during that remote era.

Seema Sahay has traced the trajectory of medicinal history in ancient India. She has identified the potentiality and dynamism of Ayurveda from a historical perspective. Diseases with their cures and remedies have always caught human attention for ages. In ancient times, people made an earnest effort to understand the dynamics of diseases, their causes and their effects. In the course of time, the understanding of the fundamentals of diseases and afflictions led to their cures and remedies as well that would ensure good health and well-being. This curiosity and eagerness led to the rise of healing traditions in ancient India which vastly contributed to the rich history of the Indian traditional system of medicine.

Melarkode Narayanaswamy Rajesh has explored the development of the ethnic system of Tibetan medicine from its Indic roots. He has highlighted the various characteristics of this folk medicine and explained the process of its transformation from Ayurveda to a hybrid system incorporating other healing traditions of ancient China and other regions. Later on, all these traits, according to the author, were synthesized under Buddhist influence having a vast collection of texts *vis-à-vis* practices in Tibet.

Arun Kumar Jha has put in harness upon the Tibetan sources to delve deeper into how the nature of the mind can be understood within the framework of Tibetan Buddhism and be appreciated by the neuro-psychobiologists as well. He emphasized the need for engaging in such an investigation that is essential to learning how to transform the mind through apt forms of meditation. These would finally facilitate to perfectly realized human souls. This, in fact, is a resplendent path of love and compassion.

The presentations in the volume expect both the readers and the researchers and also the laymen who are keenly interested in comprehending the study of the vivid vista of science and technology in ancient India.

Arun Kumar Jha
Editor

Notes

1 Pt. Shri Ram Sharma Acharya, *Rg Veda*, Part 1, Sanskriti Sansthan, Bareilly, 1965, *Nasadiya Sukta*.10.129.
2 Romila Thapar, *Cultural Pasts: Essays in Early Indian History*, Oxford India Paperbacks, 2004, p. 52.
3 Sukla Das, "Institutional Care for the Sick: Concept of Hospital in Early India", in Nupur Dasgupta and Amit Bhattacharyya (eds.), *Essays in History of Science, Technology and Medicine*, Setu Prakashani, Kolkata, 2014, p. 251.
4 Samir Amin, *Eurocentrism*, (tr.) Russell Moore and James Membrez, NYU Press, USA, 1989, *Preface*, p. vii.
5 M. S. Khan, "'Qadi Said-al-Audalusi' Account of Ancient Indian Sciences and Culture", *Journal of the Pakistan Historical Society*, Vol. XLI, Pakistan Historical Society (patronized by the Hamdard Foundation, Pakistan), Karachi, 1997, pp. 1–31.
6 Abdur Rahman (ed.), *Science and Technology in Indian Culture: A Historical Perspective*, NISTADS, New Delhi, 1984, pp. 22–23.
7 Deepak Kumar, *HISTEM in South Asia: An Introduction*; this address builds upon (the author's) presentations at the Universities of Calicut, Chandigarh and Sambalpur.
8 Quoted in Nupur Dasgupta and Amit Bhattacharyya, *op. cit.* Introduction, p. ii.
9 Scientific information is distributed in the Sanskrit, Pali and Prakrit literatures. Some of the facts as gleaned through these scriptures are intermixed with spiritual notes.
10 For a detailed study, see Sukla Das, "Perception of Life in Twilight Years in the Light of Sushruta Samhita", in Keka Dutta Roy and Chitta Ranjan Mishra (eds.), *Reflections in History: Essays in Honour of Professor Amalendu De*, Raktakarabee Publisher, Kolkata, 2009, p. 680.
11 R. S. Bisht, *Harappan Hydro Engineering and Water Management*, Lectures delivered at International Centre for Theoretical Sciences, TIFR, Mini Workshop on *Future of the Past*, on 26 November 2011 at Gateway Hotel, Mangalore; Tony Joseph, *Early Indians*, Juggernaut Books, New Delhi, 2018, p. 111; also see M. K. Dhavalikar, *Environment and Culture: A Historical Perspective*, Bhandarkar Oriental Research Institute, Pune, 2002, pp. 67–72.
12 M. N. Deshpande, "Archaeological Sources for the Reconstruction of the History of Science of India", *Indian Journal of History of Science*, Vol. 6, No. 1, INSA, New Delhi, 1971, p. 9.
13 Dilip K. Chakrabarti, *INDIA: An Archaeological History*, OUP, New Delhi, 1999, pp. 164–190.
14 For details, see O. P. Jaggi, *Dawn of Indian Technology*, Atma Ram & Sons, Delhi, 1969, pp. 106–123.
15 *Sulvasutra* literally means 'rules of the cord' used for measuring sacrificial altars.
16 Bibhutibhushan Datta, *The Science of the Sulba*, University of Calcutta, Calcutta, 1932, pp. 180–188.
17 N. Gopalakrishnan, *Indian Scientific Heritage*, Indian Institute of Scientific Heritage, Trivandrum, 2000, pp. 126–137; also see Dhruba Mukhopadhyay (President, B.S.S.), *Science in Ancient India - Reality vs Myth*, Breakthrough Science Society, Kolkata, 2016, pp. 16–18.

18 *Ibid.*, p. 134.
19 Early Indian Cosmology has been well explained by C. P. S. Menon in his book *Early Astronomy and Cosmology: A Reconstruction of the Earliest Cosmic System*, George Allen and Unwin Ltd., London, 1932, p. 94.
20 Kenneth G. Zysk, *Religious Medicine: The History and Evolution of Indian Medicine*, Routledge, 1993, pp. x–xii.
21 R. S. Sharma, *India's Ancient Past*, OUP, New Delhi, 2005, p. 311.
22 A. L. Basham, *The Wonder that was India*, Picador, London, 2004, p. 493. For details, see Hugh Thurston (ed.), "The Astronomy of Aryabhata", in *Early Astronomy*, Springer, 1996, pp. 178–189; also see Christopher Walker (ed.), *Astronomy before the Telescope*, British Museum Press, London, 1996, pp. 127–129.
23 P. V. Kane, *Journal of Royal Asiatic Society Bulletin*, I, 28:159, The Asiatic Society, Bombay, 1953, pp. 159–160.
24 Irfan Habib, *Medieval India: The Study of a Civilization*, National Book Trust, New Delhi, 2008, pp. 19–20; for a summary of the evidence on Indian decimal notation, see Irfan Habib, "Joseph Needham and the History of Indian Technology", *Indian Journal of History of Science*, Vol. 35, No. 3, INSA, New Delhi, 2000, pp. 271–274.
25 Rusdin Muhalling, et al., "Hisab and Ru'yah in Islamic Law (Syara')", *International Journal of Humanities and Social Science*, Vol. 6, No. 11, Centre for Promoting Ideas, USA, 2016, p. 306.
26 John Francis, *Philosophy of Mathematics*, Global Vision Publishing House, New Delhi, 2008, p. 38.
27 G. A. L. Sarton, *Introduction to the History of Science*, Vol. I, From Homer to Omar Khayyam, Robert E. Krieger Publishing Company, Huntington, NY, 1975, pp. 450, 475, 513.
28 B. S. Yadav and Man Mohan (eds.), *Ancient Indian Leaps into Mathematics*, Birkhauser, Boston Inc., Secaucus, NJ, 2010, p. 88.
29 For a comprehensive study on *Aryabhatiya*, see David Edurin Pingree, "History of Mathematical Astronomy in India", in Charles Coulston Gillispie (Chief Editor), *Dictionary of Scientific Biography*, Vol. 15, Under the auspicious of American Council of Learned Societies, New York, 1978, p. 533; see also P. C. Sengupta, *Aryabhata, the Father of Indian Epicyclic Astronomy*, Reprint, in Debiprasad Chattopahyaya (ed.), *Studies in the History of Science in India*, Vol. II, New Delhi, 1982, pp. 537–590.
30 *Ibid.*; also see Walter Eugene Clark (tr.), *The Aryabhatiya of Aryabhata: An Ancient Indian Work on Mathematics and Astronomy*, The University of Chicago Press, Chicago, IL, 1930, *Gola* 2 and *Gola* 5.
31 A. L. Basham, *op. cit.*, p. 498; also see P. Jha, *Aryabhata-I and the Value of π*, quoted in, J. J. O'Connor and E. F. Robertson, *Aryabhata the Elder*, University of St. Andrews, Scotland, 2000, pp. 54–59.
32 S. M. R. Ansari, "Aryabhata I, His Life and His Contributions", in D. J. Saikia (ed.), *Bulletin of the Astronomical Society of India*, Vol. 5, No. 1, Hyderabad, 1977, pp. 10–18.
33 George Abraham and Helaina Selin (eds.), *Encyclopaedia of the History of Science, Technology and Medicine in Non-Western Cultures*, 1, Springer Science, 2008, pp. 1035–1037.
34 Revd. E. Burgess, *Translation of the Surya Siddhanta*, (eds.) P. Gangooly and P. Sengupta, University of Calcutta, Calcutta, 1935, Ch. II–Ch. III.
35 Arthur Berriedale Keith, *A History of Sanskrit Literature*, Clarendon Press, Oxford, 1928, p. 524.
36 George Gheverghese Joseph, *Indian Mathematics: Engaging with the World from Ancient to Modern Times*, World Scientific Publishing Europe Ltd., London,

2016, p. 232; also see B. S. Jain, "On the Ganita-Sara-Samgraha of Mahavira (c. 850 A.D)", *Indian Journal of History of Science*, Vol. 12, No. 1, INSA, New Delhi, 1977, pp. 17–32.

37 K. S. Shukla (ed.), *Patiganita of Sridharacharya*, Lucknow University, Lucknow, 1959, pp. vi, xv.

38 Kim Plofker, *Mathematics in India*, Princeton University Press, Princeton, NJ, 2009, p. 158; also see Takao Hayashi, *Bakshali Manuscript*, in George Abraham and Helaine Selin (eds.), *op. cit.*, pp. B1–B3.

39 H. T. Colebrooke (tr.), *Algebra, with Arithmetic and Mensuration from the Sanskrit of Brahmagupta and Bhaskara*, John Murray, Albemarle Street, London, 1817, p. 156.

40 Henry James Jacques Winter, *Eastern Science, An Outline of Its Scope and Contribution*, Reprint, John Murray, London, 1952, Ch. III, for a detailed study of solving the Pellian equations $N \times Z + 1 = y^2$, $N \times Z + C = y^2$, see S. N. Sen, in D. M. Bose, S. N. Sen and B. Y. Subbarayappa (eds.), *A Concise History of Science in India*, INSA, New Delhi, 1971, p. 169.

41 Irfan Habib, *op. cit.*, p. 21.

42 H. J. J. Winter and A. Mirza, *Journal of Asiatic Society of Bengal*, Vol. 18, No. 1, Calcutta, 1952, pp. 1–10.

43 A. L. Basham, "The Practice of Medicine in Ancient and Medieval India", in Charles Leslie (ed.), *Asian Medical Systems*, University of California Press, Berkeley, CA, 1976, pp. 18–43.

44 R. S. Sharma, *op. cit.*, p. 313; for details, see Vaidya Bhagwan Das, *Materia Medica of Ayurveda*, Based on Madanapala's *Nighantu*, B. Jain Publishers Pvt., Ltd., New Delhi, 2001, Ch. 1–Ch. 9, Ch. 12.

45 Dominik Wujastyk, *The Roots of Ayurveda: Selections from Sanskrit Medical Writings*, Penguin Books, 2003, pp. 149–551.

46 *Ibid.*

47 G. K. Nariman, M. Winternitz, S. Levi and E. Huber, *Literary History of Sanskrit Buddhism*, Reprint, Motilal Banarsidass, New Delhi, 1972, pp. 276–278.

48 Jeannette Mirsky, *Sir Aural Stein: Archaeological Explorer*, University of Chicago Press, USA, 1998, pp. 81–82.

49 A. L. Basham, *The Wonder that was India*, *op. cit.*, pp. 501–502.

50 H. R. Zimmer, *Hindu Medicine* (H. N. Lectures), John Hopkins Press, Baltimore, MD, 1948, p. 82.

51 J. Jolly, *Physicians and Therapy*, (tr.), in Debiprasad Chattopadhyaya, *op. cit.*, Vol. I, pp. 175–208; also see Irfan Habib, *op. cit.*, p. 22.

52 Jean Filliozat, *The Classical Doctrine of Indian Medicine: Its Origins and Its Greek Parallels* (tr.), Devraj Channana, Munshiram Manoharlal, Delhi, 1964, p. 124; also see D. M. Bose, S. N. Sen and B. Y. Subbarayappa, *op. cit.*, p. 314; see also Dhruba Mukhopadhyay, *op. cit.*, pp. 35–36.

53 A. L. Basham, *op. cit.*, pp. 499–500.

54 Frances Garrett, *Religion, Medicine and the Human Embryo in Tibet*, Routledge, 2008, pp. 23–32; also see Terry Clifford, *Tibetan Buddhist Medicine and Psychiatry – The Diamond Healing*, Samuel Weiser, Inc., York Beach, Maine, USA, 1984, pp. 53–55.

55 C. P. Kala, "Health Traditions of Buddhist Community and Role of Amchis in Trans-Himalayan Region India", *Current Science*, Vol. 89, No. 8, Bangalore, 2005, pp. 1331–1338.

56 Tom Dummer, *Tibetan Medicine and Other Holistic Health-Care Systems*, Paljor Publications, New Delhi, 2001, p. 4.

1 Perspectives on Scientific Knowledge in Ancient India

Deepak Kumar

etat sarvam yukti-mulam, eva na tva agama-mulam
(All this is rooted in reason, not in the dogmas.)

– Neelkantha (AD 1444–1545)

Distant past always fascinates and often tends to border on romanticism. Yet its reconstruction must be exceptionally difficult; the paucity of sources, knowledge of classical or almost extinct languages, problems in textual interpretation and the necessity to contextualize, all these make it a formidable exercise. How does one place scientific and technological developments in our civilizational growth?[1] What constitutes science in the context of our ancient past? How much of it is metaphysics or religion? Did theory and praxis differ? How does one infer the nature, culture and society links? How do the Indian scientific and technological traditions appear in comparison with similar traditions in other cultural areas and societies? An enormous amount of work has been done in these areas, and as a scholar not trained in ancient Indian history and its tools, I can only recapitulate from published works with some questions that arouse my curiosity.

Every society, however crude it might appear from the benefit of hindsight, has a certain amount of scientific rationality and endeavour within it. Science is intrinsic to human nature. So has been man's fascination for tools as *home faber*. South Asian society, from time immemorial, has nurtured a thinking civilization. It never lived an isolated existence and never displayed xenophobic tendencies, continuously evolving as a result of each politico-cultural interaction with the outside world and social change within the region. In pre-modern times, South Asia was known for its contribution to astronomy, medicine and mathematics. In ancient India, medical men, in spite of their scripture orientation, insisted on the supreme importance of direct observation of natural phenomena and the technique of rational

DOI: 10.4324/9781003370642-2

processing of empirical data. The *Charaka Samhita*, an ancient medical text, says, "To one who understand, knowledge of nature and love of humanity are not two things but one".[2]

Unlike the Cartesian worldview, in India, nature and culture, and subject and object were never seen in an adversarial mould. Pluralism was recognized, but at the core there remained a quest for synthesis and a holistic understanding. This trend is manifest especially in the Indian medical thought and system. Again, it is this trend which takes even "*yukti-vyapasthraya bheshaja*" (rational medicine) into the mirth and mire of metaphysics. How does it take place? From magico-religious beginnings to rationalistic therapeutics and then back to decadence – it must be a fascinating journey. Or, is it possible to think of another trajectory? If not, what were the factors responsible for the ups and downs? Was it because of the priestly class and the "*varnashrama*" hierarchy? The priests would naturally have been interested in supernaturalism and the mystification of nature. Why did the medical men bow to them? Were there other options? Linear, hagiographical accounts or even philosophical explanations do not suffice. We need critical sociological enquiries, as shown by Debiprasad Chattopadhyaya.[3]

Here, one may like to pause to reflect upon the nature and place of knowledge in Indian traditions. Ancient Greeks perceived knowledge in terms of power; it was used to acquire and consolidate power; it was not merely knowledge for the sake of knowledge. To Indians, it meant liberation – *sa vidya ya vimuktaye*. But liberation from what: ignorance, disease, poverty or what Buddha called *dukkha* or liberation from this world itself? As the Brahmanical tradition became strong, the metaphysical other-worldliness overtook the quest for *vimukti*. The purpose may have changed from time to time, but the quest remained. It is this quest that makes our civilizational journey fascinating. Inherent here is also the premium on wisdom. Wisdom was the aim of our rishis, and wisdom was applied even for developing techniques to improve the quality of life, for instance, Ayurveda or Yoga.[4]

The Indian knowledge system may not have what Kuhn terms "paradigm shifts". The change was there but evolved slowly; knowledge did move forward but without any sharp break with the past. It moved through writing commentaries which scholars in every generation wrote making additions or subtractions to the then-existing corpus of knowledge. Significantly, *yukti* (skill, reasoning) is highly prized in the Indian tradition. Neelkantha praises his hero, Aryabhata, as *yukti-nidhi*, and Bhaskara II praises his *bija-ganita* as *avyakta-yukti* (handling unknown variables). Unlike Greek "axiomatism", the Indians used "computational positivism". This attitude relied only on facts as the possible objects of knowledge; it did not give any credence to metaphysics or to assumptions that could not be verified as facts. The result was that errors in the predictions of planets were less in the Indian approach compared to those based on Ptolemy's "axiomatic" procedure. Another interesting aspect is that the Indians considered the Siddhantas valid only for some finite period of time (*kvachit kale pramanam*) and not valid forever.

The biggest problem in Indian tradition is that of "*praksipta*", which are basically additions reflecting the culture and practice of the time. The "*praksipta*" merges well with the text and in most cases difficult to identify. In fact, in the distant past, knowledge advanced through additions and subtractions in the form of commentaries (for example, *tika*, *tippani*, *bhasya*, *nighantu*). Most of the additions were the results of thorough deliberations and relevant in view of the then-existing knowledge. However, there were many which were not only untrue but basically selfish.

In recent years, environmental issues have attracted the attention of everyone. It is no doubt an extremely important area of research requiring an inter-disciplinary knowledge involving geography, geology, chemistry, zoology, medicine, anthropology and history. We now have highly specialized studies on archeo-zoology, palaeontology, palynology (plant fossils), dendrochronology (the sequence of tree rings), etc. But we need to guard against environmental determinism. For example, some argue that the Indus Civilization collapsed because of growing aridity in the Indus basin. Similarly, the influence of meteorological phenomena like El Niño is held responsible for numerous famines or other extreme situations. Yet, it is true that certain environmental "pressures" might have helped migration of humans and animals or may also have induced introduction of certain wild cereals or caused some specific "miasmatic" diseases.

Environmental understanding of our ancient past begs several questions. For example, is there an Asian perception of nature or science, and is there a western perception of science or nature which can be applied universally? Are there cross-cultural differences? How and what do we learn from them? What are the similarities and dissimilarities between a society's prevalent perception of nature and its actual human behaviour? There have always been differences in perception versus behaviour. Do we in some Asian cosmologies find only an elastic, relativistic naturalism, linking the physical environment to the demands of society rather than dichotomizing nature and culture? [5]

In the Indian worldview, nature and culture, and subject and object were never seen in an adversarial mould. But it was definitely complex; different thought processes and views were expressed and these could co-exist. So, you have superstition and reason both together. Again, it is this trend which takes even *yuktivyapasthraya bheshaja* (rational medicine) to the mirth and mire of metaphysics.[6]

Split Science

For the ancient periods, we should not use the term science as we understand it today. Scientific investigations were definitely there; empirical observations were made, and numerous rational explanations were given. But it is doubtful whether the methods used were as exact as done in modern times. In ancient texts, one finds numerous references to *yukti* (reason), *tarka vidya* (science of reasoning), *anvikshki* (investigation), *anumana* (intelligent inference), *pramana* (proof) and *hetusastra* (dialectics). However, the Indian traditions did accommodate certain rational explanations, but taken as a whole, more reliance was placed on *anubhuti* (experience) than on anything else.

The spiritual experience thus came to form the bedrock of our religion. The great philosopher Sankara preferred *anubhuti*, not *suska tarka* (dry reason). Even today, all men of religion put a premium on their sacred texts and their "experience", and not on reason alone. This "experience" along with creative imagination and some rationale was codified into sacred texts which gradually acquired canonical status. To question or challenge a canonical text or knowledge (*sastra-sammat jnan*) has never been easy in any society. Many perished in the process, and a vast majority had to compromise. Brahmagupta, the author of *Brahmasphuta Siddhanta* (AD 628), knew the real reason for the eclipses, but since the Puranas had claimed that it happened because of the demon Rahu, he preferred to save his skin by endorsing the Pauranic claim. He took one step forward and then two steps backwards. The power of tradition and the weight of canonical knowledge were so great that even great minds had to succumb.

Why did scientists behave like split personalities? Why, while accepting science, they also prostrate themselves before its opposite? Why do different kinds of alien ideas and attitudes remain mixed up with science in the ancient texts? It was because the lawgivers, who guided and guarded their traditions jealously, were not willing to allow any challenge to the authority and their worldview. So, as a result, we get a knowledge corpus wherein, as Al-Beruni rightly puts it, "gems were mixed with cow dung"! Debiprasad Chattopadhyaya argues that in ancient India, the only discipline that contains the clear potential for the modern understanding of natural science is medicine. The other disciplines like phonetics, grammar (*vyakarna*), etymology (*nirukta*), metrics (*chandas*) and calendrical astronomy (*jyotis*a) had elements more of rituals and orthodoxy than those of rational observation and explanation. Medicine too had a magico-religious origin, but gradually it developed from *daiva-vyapasraya bheshaja* to *yukti-vypasthraya bheshaja*. Because of this development, and also because the ancient physicians wanted to gain autonomy and independence, they elicited criticism and condemnation by all lawgivers from *Apastamba o Manu*. Buddha appears as a dissenter and gives medical men a place of honour. Later, he was himself hailed as a healer par excellence, much like Jesus, but without magic at his command.

Ayurveda gradually emphasized not only direct observation of natural phenomena but also the technique of rational processing of empirical data.[7] Their understanding of the aetiology of several diseases and even the functioning of the organs of the body might have been deficient, but their knowledge of the therapeutic value of the plant world was amazing. It is being worked upon in modern times too. It is interesting to see how the ancient medical men try to balance the contradictory pulls and pressures. The *Charaka Samhita* begins with respect to the gods, cows and the Brahmanas, but later it goes on to describe the medicinal value of beef in certain kinds of grave illness (*visam jvar*). Similarly, it extols the virtues of celibacy (*brahmcharya*) and then goes on to describe recipes for increasing sexual stamina and encouraging more sex in the winter months. The text also describes the technique of preparing 84 varieties of alcohol. In fact, what looks like a contradiction is basically a statement of truth. The Upanishads identify the *purusa* with soul and the Ultimate Reality (*Brahmana*). But, Charaka in his *Samhita* makes the *purusa* material and physical (*sarir-mulak*) and does away with the spiritual connotation. In pregnancy, the foetus is described as the transformation of matter. The physical world and nature itself are taken as inextricably interlinked. Sushruta talks of mutual interrelation (*paraspara samsaragat*), mutual interdependence (*paraspara-anugrahanat*) and mutual interpenetration (*paraspara-anuprevesat*). This is how the panchbhutas (Five Forms of Matter) run the universe.[8]

Against this background, the ancient Indian physicians developed a theory of medicine revolving around *tridosa* (three humors) and claimed that the diseases occurred because of the imbalance among the three. This knowledge helped the physicians (vaidyas) restore the balance concerning the three humors. He was supposed to work with his hand and also with instruments when needed. Sushruta lists more than a hundred surgical instruments. Work with hand signified dignity of labour. The physician was supposed to know the qualities (*guna*) of substances (*dravya*) to be used for treatment. They were expected to understand the nature (*svabhava*) of the disease, the patient and the drugs. Nothing was to be left to the unknown or invisible (*adrasta*) or to the *karma*. The medical situation was thus seen as an interaction between environmental matter and body matter. The effect of nature (*svabhava*) was to be seen in the condition and work of the three dosas (humors). This indeed was a great limitation on their knowledge because not everything could be explained in terms of the *pancha-bhuta* and the *tridosa* framework. This also explains the weaknesses of the ancient Greek and Indian medical systems, which much later were successfully challenged in the early modern times by scientists like Paracelsus, Vesalius and Harvey.

In India, for centuries canonical texts like the Samhitas of Charaka and Sushruta were there to guide, but knowledge advanced through commentaries and even replications. In the deep south, the Siddhas had developed knowledge of pulse and methods of diagnosis by eight kinds of clinical examination. By the end of the 13th century, thanks to the works of Chakrapanidata, Vangasena and Sarangdhara, a new change had come in the shape of the blending of Ayurveda and *Rasashastra* (medicine and alchemy). These developments have been mostly internal, but the possibilities of external influences from other cultures cannot be ruled out. Unfortunately, these texts and later commentaries have no anatomical or surgical illustration.

Dissection became a taboo and knowledge became secretive, confined to one's own son or the Brahmanas. *Todarananda*, a 16th-century encyclopaedic work, for example, says that *rasavidya* (alchemy) is to be kept "as secret as mother's genitals". Even mineral classification reflected the caste divisions. For example, diamonds are described as of four types, *Brahmana*, *Kshatriya*, *Vaisya* and *Sudra*; they are also male, female or "*napunsaka*". The *Anandakanda* (The Root of Bliss) encompasses almost the entire Hindu alchemical tradition in 6000 verses. It is the sole attempt to fuse rasa siddhas (alchemists) with the nath siddhas (spiritual tantries/yogis).

Apart from attempts for synthesis, there may be some examples of total repudiation or departure. Dominik Wujastyk has recently brought to notice one such rare example – a text titled *Rogarogvada* (Debate on Illness and Health) composed by Viresvara in 1669.[9] This text explains systematically the principal theories of Ayurveda and then refutes them one by one. For example, the authorities define disease as identical to an inequality in the humors, and yet they accept that the humors may naturally exist in different quantities without causing illness. Phlegm predominates at the start of the day or after a meal, but one is not always ill after a meal! And so the central doctrine that humoral imbalance is identical to disease must be wrong. Thereafter, Viresvara gives a curious explanation; diseases come and go for no apparent reason. Just like the rising and setting of stars, disease is the pain of the mind, body or sense organs, and it arises for no reason. This explanation may appear quixotic, but it is nevertheless offered, as Wujastyk argues, in "a spirit of intellectual rigour and debate which speaks of an original if impulsive mind".[10]

Religion and Science

The terms religion and science are easier to say than to explain, that too in relation to the distant past. Religion and knowledge have always been intertwined and intermeshed. In India, the earliest ideas and rituals were articulated in the four Vedas. The *Veda* literally means "to know" or "knowledge". Knowledge of the natural world and the quest for the supernatural continued simultaneously. There did appear what one may call the Vedic natural laws (*rta*). Knowledge of the stars and planets was necessary for the timing of rituals and sacrifices. The altars had to be geometrically designed, and the Vedic chants were to be metered. All this required a certain amount of scientific knowledge.

The *Rigveda* states that the sun neither sets nor rises. It travels through a spacious arc way which has been interpreted as either Milky Way or elliptic. The Vedic people probably knew the solstices. They also knew that a year consists of 360 days and 12 months, with a month of 30 days divided into the light half (*sukla*) and the dark half (*krsna*). The solar eclipse was described in terms of a demon attacking the sun and its gradual liberation by the sage Atri. This liberation was described along with the change of colour in the sun's disc. In the later *Brahmana* literature, there appear two terms "*naksatradarsa*" (star-gazer) and "*ganaka*" (calculator). It seems that the observation of stars and theoretical calculations went together.[11] These calculations would not have been possible without substantial mathematical knowledge. Astronomy and mathematics were twin sciences, and astronomers were mathematicians at the same time. In the Vedic period, the *Sulvasutras* (Rules of the Measuring Chord) were the repositories of mathematical knowledge and methods. Even the origin of the Pythagorean Theorem has been traced to these sutras. Around the same time, another branch of knowledge came up known as the *Vaiseshika Sutras* which dealt with the physical concepts relating to substance, motion, space, time, etc. Unfortunately, these remained couched in a language that mostly appears incomprehensible.[12]

Numerous names appear as astronomers of the post-Vedic period. They are Garga, Parasara, Kasyapa, Narada, Vasistha, Bhrigu and many more who composed their own Samhitas. Garga is known for his comment on the Greeks whom he described as mlecchas who deserved to be honoured as sages because of their knowledge of astronomy. All this led to the development of a kind of knowledge which formed the bedrock on which new additions were to be made. These new additions came to be known as the Siddhantas, which were probably 18 in number. Most of them were fairly inaccurate, but of them, the *Suryasiddhant* was found to be the most accurate and formed the basis for the works of celebrated astronomers like Aryabhata and Varahmihira. It is significant to note here that in ancient India, knowledge moved not through paradigm shifts but through a slow and time-tested process of additions and subtractions. There were several path-breaking texts, no doubt, but every generation wrote commentaries and thereby kept the pace of knowledge gathering.[13]

It is said that by the 4th century AD, the Indian mathematicians had developed the system of decimal numeration using nine digits and zero. Aryabhata's *Aryabhatiya* is a well-known mathematics-based astronomical text composed in the 5th century AD. This work marks a paradigm shift. It deals with astronomical principles and methods of computation in a highly condensed form. Aryabhata was probably the first astronomer to talk about the rotation of the earth on its axis and also provide the correct explanation for eclipses. He was followed by Varahmihira, Bhaskara I and Brahmagupta. The last one composed *Brahmasphutasiddhanta* consisting of 1008 verses. It dealt with planetary motions, eclipses and the problems of time, space and distance. He criticized Aryabhata for abandoning the traditional Rahu-Ketu theory of the eclipses. Here is an example of how an astronomer had to accept something handed down by the religious tradition epitomized by the Puranas. But he was a great mathematician and his works were translated by the Arabs which started a new trend of mathematics-based astronomy. The tradition continued and another landmark appeared in the 12th century AD in the works of Bhaskara II. His

Siddhantashiromani still retains its pre-eminent position among astronomers. He dealt with arithmetic, algebra (*bijaganita*), knowledge of spherics (*goladhyaya*) and astronomical instruments (*yantradhyaya*). Scholars agree that despite the scriptural orientation and being couched in religious terms, the astro-mathematical knowledge of ancient India represents a thinking society.[14]

The Buddhist and the Jain Interventions

The emergence of Buddhist philosophy marks a watershed in ancient Indian thoughts. Unlike contemporary *Brahmana* literature, Buddhism puts great emphasis on reason and rationality, and like science, it tries to explore the realms of sensory and mental experience. It brings the mind and body together. Buddhism does not define itself as a religion or science, nor does it distinguish between religious truths and scientific truths. But both Buddhism and science teach that all beings are alike subject to universal laws. Buddhism examines the laws of the inner world, while science explains the outer world. Some scholars describe the early Pali texts as "an analytical approach, combined with empiricism". They take the theory of *madhyamaka* as part of not only the philosophy of Buddhism but also the philosophy of science. They argue, "Science deals with matter, Buddhism with mind. Science is rationalist, Buddhism experiential. Science is quantitative, Buddhism qualitative. Science is conventional, Buddhism contemplative. Science advances humans materially, Buddhism spiritually."[15]

From the psychological point of view, it is significant that the Buddhist traditions believe that consciousness (like *vijnana*) is not static; it is dynamic and moves from one life to the next. Cartesian dualism looks at "solid" matter and "nonmaterial" mind in a position to each other. But for Buddhism, this conflict between materialist and idealist points of view (i.e. between mind and matter) is simply false. It denies such quality and rather argues that neither consciousness nor the material world has any intrinsic reality. Modern physics holds that mass–energy would never completely disappear in the material world – "nothing is lost, nothing is created". Buddhism applies similar reasoning to consciousness; it has no beginning and no end. In place of duality, this philosophy emphasizes the interdependence between our world and the consciousness perceiving it. This harmony constitutes the kernel of the "Middle Path". In Buddhism, scientific and contemplative knowledge are not antagonistic at all, though their priorities may be different.

Some scholars have compared the Buddhist *madhyamaka* with the modern quantum theory. For example, many properties of an electron are not intrinsic to the electron itself. Similarly, in *madhyamaka*, both the subjects and objects are relational and not intrinsic. A *dharma* has no *svabhava* (nature) and thus its intrinsic nature is denied. Similarly, no electron can in itself be said to be either a wave or a particle. Its description depends on what kind of experiment was being done, and how it was being observed. In Buddhism, no such contingency exists.

Another important ancient Indian religion is Jainism. It developed a set of cosmological theories of its own and differed vastly from the Brahmanical *Vedanga Jyotisha*. Its original text is *Suryaprajnapti* attributed to Lord Mahavira himself (5th century BC). Numerous commentaries appeared at a regular interval. Jain astronomy has two suns, two moons and two sets of 27 nakshatras. The earth is conceived as a series of flat concentric rings of land masses separated by concentric ocean rings. At the central circle was located the Jambudwipa at whose centre stood the magnificent Mount Meru. No wonder, such a description was found by a colonizer like Lord Macaulay as such that would elicit laughter in an English girls' boarding school! Unlike astronomy, the Jains seem to have contributed more to mathematics. A 1st-century BC text, *Sthangasutra*, refers to the science of numbers, fundamental operations, solid mensuration, fractions, algebra, permutations and combinations, etc. They excelled in the enumeration of large numbers. Scholars have compared this ability of the Jain mathematicians to that of the Alef-zero of modern mathematics.

Technologies

Tool making has been synonymous with civilization growth. The wheel-turned pottery was very common in the Indus Valley Civilization. The pottery was kiln-fired and painted. Mica, sand and lime were used as tempering materials, and charcoal was used as fuel. Other than glazed pottery, the Harappans knew the use of copper and also the techniques of making beads, soldering, sheet-making, riveting, coiling, casting, etc. The "dancing girl" of Mohenjodaro is a remarkable evidence of metal craftsmanship. Weapons of warfare were also made out of copper–bronze combination. Arrowheads were also made of metals. Iron, however, comes a little later. But when it came, it revolutionized agriculture and the economic life of the people. The Vedic people were good metalworkers.

Iron was known in the Middle Eastern Civilization. Probably came to India by the 5th century BC, Indians had begun to excel in iron smelting. They could produce a kind of steel, also known as wootz, which was exported and used in the famous Damascus sword. The metalworking reached its height during the period, and its finest example still stands erect in Delhi. This iron pillar manufactured in the early 5th century AD has not rusted and shows no sign of decay. The iron used here is almost pure with high phosphorus, low sulphur and no trace of manganese.

This process involved not only a very meticulous composition of the ore but also the use of high temperatures. How they could master it is still a mystery. Casting copper did not require high temperature, but iron smelting was a different ball game. Then through pottery and copper, the Iron Age had emerged which required a superior know-how and a more efficient use of energy. This meant the production of new items and new tools and an increase in trade. The introduction of iron hoe in the plough enhanced agricultural production many fold. A new economy had emerged which continues to grow to date. The use of iron had changed the face of war as well. The celebrated *Arthasastra* gives a lot of information on mining, smelting and the utilities of different kinds of metals.

A defining feature of modern civilization is the use of glass which made the manufacture of telescopes and microscopes possible and thus opened revolutionary methods of exploration and investigation. The glass and silicates were known even to the Harappans. Silica and alkalies were the basic materials, and copper was used as a colouring agent. But it took several centuries before we learnt to grind glass and produce revolutionary changes.[16]

Another significant area in which the Indians excelled was chemical technology. They were excellent alchemists. Through experimentation, observation and inference, they could develop distillation processes dealing with lead, mercury, copper, gold, silver and several plant objects. This came to be known as *rasavidya*. It was categorized as eight maharasas, eight uparasas, the eight dhatus and so forth. So much has been written on *rasavidya* in India; almost every generation wrote several texts and commentaries on it. It assumed mythological and later tantric connotations. It came to be associated with longevity, sexual potency, in short, the elixir of life. Mercury formed the basis of such investigations and was hailed as the semen of Lord Shiva. Elaborate processes of its purification (*sodhana*) are given in numerous *rasasastra* texts. It seems its practitioners had proper laboratories (*rasasala*). The Indians had made remarkable growth in alchemy and chemistry much before Paracelsus (early 16th century AD).[17]

Plant Sciences and Agriculture

The archaeological explorations from the sites of the Indus Valley Civilization have thrown ample evidence of botanical knowledge. Wheat, barley, rice, cotton and different kinds of goods had been found in the Vedic period. There are numerous references to plants and cereals in the Samhitas and the Brahmanas. Of course, botanical knowledge had not been systematized, but the people knew about the morphology and distribution of plants. Even some categorization was attempted. For example, some plants were described as those of the spreading variety (*prastranati*), some bushy (*stambini*), some jointed (*kandini*), some with large branches (*vishakha*) and so forth. They were broadly divided into trees (*vrksha*), herbs (*osadhi*) and creepers (*virudh*). Plants appear extensively in the medical literature. For Charaka, an expert physician is one who knows the herbs botanically, pharmacologically and in every other respect. Without plant knowledge, *bheshajavidya* would be utterly incomplete. In fact, a distinct knowledge (*veda*) appeared known as *Vrksayurveda*, and Kautilya's *Arthasastra* has devoted separate sections dealing with the collection and selection of seeds, germination, grafting, cutting, manuring and also the meteorological conditions.

Later in the 1st century AD, Parasara composed a *Vrksayurveda* which gives, in a rudimentary way, an outline of plant morphology, nature and properties of soil, description and distribution of forests, functions and classification of flowers and fruits, and the details on seeds (*vija*). The text by Parasara is really unique and reflects truly the state of botanical knowledge.[18] Kautilya and Charaka gave detailed descriptions of the nature of vegetation also. The regions are described in terms of *jangala* and *anupa*. The *jangala* have open spaces and dry wilderness, while the *anupa* region would have marshy and swampy tracts. Arbori and horticulture are also referred to in many texts. *Upavanvinoda* (pleasures of gardening) are referred to in great detail. However, agriculture remained central to plant knowledge and vice versa.

The Vedic farmers knew the seeds, the methods to sow and also the use of plough. It is said that the system of sowing began in India. They also knew the method of improving the fertility of the soil through crop rotation. They knew the significance of irrigation and constructed channels for carrying from different sources. It is possible that they used wheels and earthen pots (*ghata*) to lift water. The Mauryans are known to have patronized irrigation schemes and constructed large tanks. In South India, huge tanks were constructed and some of them survived for centuries. The *Arthasastra* talks of meteorological knowledge also. Rainfall was monitored and recorded. Soil classification and land use were discussed in great detail. Land was classified in terms of cultivated (*karsa*), waste land (*usara*) and pasture (*gocara*). The *Amarkosa* describes 12 types of land in terms of their fertility and physical characteristics, and it also describes a number of agricultural implements. The treatment of seeds and manuring the soil to ensure a good crop are discussed in great detail. All this was empirical knowledge which was found useful through long periods of observation and understanding.

It may not be proper to find the similarities between modern knowledge systems in the ancient past. Relevant traces are bound to be there, but these cannot be exaggerated to the extent that all knowledge existed in the Vedas. In this regard, it may be said that we may not have everything in the ancient past, but there was definitely a substantial corpus of useful knowledge. After all, it was a thinking civilization and a vast ancient literature and archaeological evidences bear testimony to this. No wonder the celebrated author of *Tabqat-al Uman*, Qadi Saiid al-Andalusi (1029–1070), wrote,

> Among all nations, India was known as the mine of wisdom and the fountain-head of justice. Although their colour belongs to the first grade of blackness, yet God, the Exalted, has kept them immune from evil character, base conduct and low nature. He (God) has thus exalted them (Indians) over many brown and white peoples ... They also obtained profound and abundant knowledge of the movements of the stars, the secrets of the celestial sphere and all other branches of mathematical

sciences. Moreover, of all the peoples, they are the most learned in the science of medicine and well-informed about the properties of drugs and the nature of composite elements.[19]

Notes

1 D. M. Bose, "History of Science in India. How It Should Be Written?", *Science and Culture*, Vol. 29, No. 4, 1963, pp. 163–166; D. V. Subba Reddy, "Evolution of the Study and Writing of History of Medicine", *Indian Journal of History of Medicine*, Vol. VI, 1961, pp. 21–39.
2 D. P. Chattaopadhyaya, *Science and Society in Ancient India*, Research India Publication, Calcutta, 1979, p. 7.
3 *Ibid.*; see also D. P. Chattaopadhyaya, *History of Science and Technology in Ancient India: The Beginnings*, Firma KLM Pvt. Ltd., Calcutta, 1986.
4 K. G. Zysk, *Medicine in the Veda*, Motilal Banarsidas, Delhi, 1996, pp. 1–15.
5 Ole Brunn and Arnae Kalland (eds.), *Asian Perceptions of Nature*, Curzon Press, London, 1955, pp. 1–20.
6 R. C. Majumdar's enthusiastic account of *Ayurveda* is an example. See D. M. Bose, et al. (eds.), *A Concise History of Science in India*, Indian National Science Academy (INSA), New Delhi, 1971, pp. 213–262.
7 For details, see G. N. Mukhopadhyaya, *History of Indian Medicine*, 3 Vols., Calcutta, 1914, Reprint Orient Books, Delhi; Idem, 1913; *The Surgical Instruments of the Hindus*, 2 Vols., Calcutta University Press, Calcutta; P. V. Sharma (ed.), *History of Medicine in India*, Indian National Science Academy (INSA), New Delhi, 1992.
8 For a lucid and incisive account, see Dominik Wujastyk, *The Roots of Ayurveda*, Penguin, New Delhi, 1998, pp. 1–38.
9 Dominik Wujastyk, "Indian Medical Thought on the Eve of Colonialism", *IIAS Newsletter*, 2003, p. 21.
10 *Ibid.*
11 K. D. Abhyanakar, "A Search for the Earliest Vedic Calendar", *Indian Journal of History of Science*, Vol. 28, 1993, p. 1.
12 S. G. Dani, "Vedic Mathematics: Myth and Reality", *Economic and Political Weekly*, 31 July, 1993, pp. 1577–1580.
13 Frits Staal, *Concepts of Science in Europe and Asia*, IIAS, Leiden, 1993, p. 26.
14 For details, see S. N. Sen and K. S. Shukla, *History of Astronomy in India*, Indian National Science Academy (INSA), New Delhi, 1985.
15 For an in-depth discussion, see B. A. Wallace, *Buddhism & Science*, Motilal Banarasidass, Delhi, 2003.
16 H. C. Bhardwaj, *Aspects of Ancient Indian Technology*, Motilal Banarasidass, Delhi, 1979, pp. 1–5.
17 Refer to, P. C. Ray (ed.), *History of Chemistry in Ancient and Medieval India*, Asiatic Society, Calcutta, 1956; B. V. Subbarayappa (ed.), *Chemistry and Chemical Techniques in India*, Oxford University Press, New Delhi, 1997.
18 G. P. Majumdar, *The History of Botany and Allied Sciences in Ancient India*, Archives Internationales d' Historire des Sciences, UNESCO, Paris, 1951.
19 M. S. Khan, "Qadi Said-al-Andalusi's Account of Ancient Indian Sciences and Culture", *Journal of the Pakistan Historical Society*, Vol. XLV, 1997, pp. 1–31.

2 Some Methodological Questions on the Study of History of Science and P.C. Ray's History of Hindu Chemistry

Arun Bandhopadhyay

Approaches to the study of history of science underwent critical changes in the last two centuries before they could more definitively tilt towards the social dimensions of the making of science. The separate approaches indicate certain moments of these changes, but they do not exhaust all moments or aspects. The development of history of science during the last 50 years has been marked by a proliferation of methods and perspectives, rather than by the emergence of a consensus. Helge Kragh has identified at least six usual grounds of justification for the study of history of science but is fully satisfied with none.[1] These grounds require some paraphrasing, and we shall do that before coming to the Indian problem.

Firstly, it is often stated, especially by scientists–historians of history of science, that the subject, when properly studied, may have a beneficial influence on the study of science today. A slightly different version of the argument is the assertion that history of science should function as an analytical instrument for the critical evaluation of methods and concepts that appear in modern science. But none of these arguments gives credence to the view that there is a direct scientific relevance to the history of science. The relationship is no better than a sporadic influence like the one done by literature or religion.

Second, it has also been argued that history of science has three separate aims: (a) it provides materials for a critical self-examination of science, (b) it bridges the gap between science and humanities and (c) it helps us to know the genesis of scientific theories that scientists are primarily concerned with. It is stated that the increasing interest in history of science in Western countries was caused largely by the fact that the science-based technology of the war attracted to itself. It is true that history of science can be used and ought to be used to analyze the interplay between science, technology and society. But empirical evidence that science always results in technology is weak; it is also true that science and technology invariably contribute to people's welfare and safety.

Third, history of science is stated to function as a background for and in close relationship with other fields, such as philosophy and sociology of science. The links between philosophy of science and history of science have

DOI: 10.4324/9781003370642-3

become steadily stronger in recent years. However, the relationship between history of science and philosophy is complex and far from fully settled. The same is true about the relationship of history of science with the sociology of science and theory of science. Some scholars prefer to regard history of science as one element in a composite research programme, "science of science", that includes all studies relating to science in an interdisciplinary way. However, G. Krober has modified this argument by giving history of science a more independent standing, "It would be completely erroneous to talk of a simple incorporation of history of science writing under science of science or to imagine that it should be relegated to being a servant of science of science". [2]

Fourth, history of science may play a significant didactic role in demonstrating the true nature of scientific knowledge. Here, the use of history is justified by its capacity to present the sciences in a "softer" way to make them more attractive to a larger audience. History of science can contribute towards a less dogmatic conception of science and scientific methods, thus acting as an antidote to orthodoxy and uncritical enthusiasm for science. But this argument cannot be stretched too far. The history of science can also be used to support dogmas and strengthen scientific authority. In short, generally speaking, the question of didactic importance is problematic.

Fifth, history of science is often stated to reflect the humanist placement of science. G. Sarton, the famous historian of science, reminded us of the connections and shared roots of science and humanities in the 1930s and 1940s,[3] while C. P. Snow wanted to bridge the schism between two cultures in the 1950s and 1960s.[4] G. T. Clark went so far as to declare that "the history of science is, in fact and indeed, the new humanism for our contemporary, irreversibly technological and the present moment beleaguered culture".[5] However, such arguments have been critically examined. The connectivity arguments should not be used ideologically to silence the critics of contemporary science. Ironically enough, it has even been stated that the more relevant question is why the greater part of science today is no longer regarded as an expression of humanist endeavour?

Finally, there is a view that history of science does not require any pragmatic justification. As an important factor in the general social, economic and cultural development, science attracts historical attention in the same way as, for example, do religion and economic conditions. History of science has no particular aim apart from that of revealing the past. These lines of justification may typically appeal to historians, but necessarily to scientists. As I. B. Cohen wrote in 1961,

> At the present time, it is no longer necessary to justify the study of history of science ... Not far off is the time when historians of science will be so numerous that they may produce scholarly works which need satisfy only the members of their own profession, the only requirement being that of high standards.[6]

But in the process of development of history of science, occasions arise where the scientist studies nature and the historian studies the scientist.

Like the proverbial fable of four blind men visualizing an elephant, each of these justifications forms a significant part of the aims of history of science, but none of them binds the discipline as a whole. Historians try to cope with infinity, integrating their subjects with the subjects and methods of other related disciplines. New perspectives, stimulated by social and economic history in particular, have been encompassed in the history of science. Nations, firms, political agencies, institutes, individuals and scientific societies are being studied by a rising group of historians in the history of science. Indeed, whatever science or subject is being studied, history of science is regarded by its practitioners increasingly as a field of history rather than of science. The other lesson from history of science is that the positivist belief in a value-free, culturally independent science is largely a myth.

Given this general background of the intellectual tradition of writing the history of science, the discussions revolve around a few thematic contexts: the East–West encounter, the diffusion theory, the colonial impact, the nationalist and the post-colonial response and globalization.

The East–West encounter is a long-drawn-out story, the different meanings of which can be deciphered in the late pre-colonial, colonial and post-colonial eras. Dhruv Raina has elsewhere identified two cardinal features of the encounter: "how the sciences of the Orient were represented in the discourse produced by the Occident, but more importantly, how the Indian scientists and historians of science engaged with sciences of India".[7] In the short-term periodic context, the picture is sketched in terms of the domination of political and intellectual power, as in the case of the twilight of the late pre-colonial era or the high noon of the colonial era. But from the long-term point of view, a more interesting picture emerges. As Raina says,

> From the Orientalist period, through the nationalist period to the period of Nehruvian socialism in India, the forces of decolonization sought to challenge a hegemonic conception of the history of science, such that history of sciences of the non-west would not be subordinate texts to the mainstream discourse of the history of Western science.[8]

Such a change, based on a critique of "the idea of science as a cultural universal", came into being only after the end of the 1960s.[9]

The diffusion theory of science, indirectly connected with the theory of civilizing mission, has been in vogue amongst analysts for a long time. Its connotations are manifold. The initiative is invariably located in the West, while the chequered response – with all its potentialities, hesitation, partial fulfilment of promises and innate obstacles – is to be found in the East. Science is projected as the flag bearer of the civilizing mission of the West, dissipating superstition and installing new technological systems over the old, primitive ones. Britain's changing role in the international economy led the colonial state in India to be innovative in founding technical institutions and colleges in the late 19th century. Colonial encounters, through a sort of diffusion of models and practices brought from home, contributed to the development of technical education in India.

> The several perspectives emerging argue, despite differences, that the standard tale of assimilation of modern science as a Western cultural import was inadequate and missed out the multifarious nature of exchange between the modern science and the so-called traditional knowledge forms.[10]

The criticism of the diffusionist model brings us closer to the arguments pertaining to the colonial impact. Three aspects of these arguments demand our immediate attention. First, it focuses on the intimate relationship of scientific and technological projects (like the railways) to the 19th-century imperialism. Second, there was an ostensible impact of colonial rule on indigenous scientific knowledge and institutions. Third, the role of British and Indian scientists in the creation of scientific knowledge and institutions of science and engineering is to be critically examined. Joseph Needham has categorically pointed out that scientific innovations are intricately enmeshed in socio-economic formulations.[11] Zaheer Baber has even pointed out that though the Indian scientists did not respond uniformly to Western science and technology, there was a group "that was urging for heavy industrialization was dominant and its views resonated with the dominant factions of the Indian National Congress".[12]

The colonial impact is often matched by the nationalist response in the case of India. In constructing a national identity, a tradition is invented by choosing from the past which then becomes instrumental in creating the required identity. Romila Thapar thinks that historical reconstruction is central as images are drawn from the past. She adds, "If the history of science is tied to nationalist ideology, and it may be nationalism of any kind, it cannot be treated as universal. Categories such as Hindu science and Islamic science disallow universality to science".[13]

However, the nationalist discourse in India is far more complicated than is presented by Professor Thapar. There was latent and explicit racism; there were colonial reservations regarding the abilities of the Indians as scientists, and over and above all, there was competition from various quarters. It has even been argued that the study of J. C. Bose and Ramanujan sought to impress upon the people how traditional cognitive orders creatively enriched the practice of contemporary science, thus abandoning the epistemic privilege hitherto accorded to modern/Western science.[14] Thapar is also probably aware of the pitfalls of "universal science". She, therefore, hastens to add, "What we are now experiencing is something which claims to be beyond nationalism and calls itself for trans-nationalism – synonym for globalization, enriching us all to the siren of market".[15]

It is in this general context of methodological questions on the study of history of science that I wish to place some 20th-century study of history of science in ancient India. It is argued here that the historical placement of the study is most important and the standard questions like the East–West encounter, the diffusion theory, the colonial impact and the nationalist discourse – the questions that became very important in our understanding of history of science in India in the 20th century – lay at the root of the publication and analysis of P. C. Ray's celebrated work on *History of Hindu Chemistry* in two volumes (London and Calcutta, 1903–1909). This is apparent not only from the preface and forward of the study but also from the very textual analysis of the subject as done by Professor Ray.

What was there in the preface of the first edition of *History of Hindu Chemistry*, Volume I,[16] that sounds most important from the methodological point of view? Professor P. C. Ray has no doubt that such studies are the product of orientalism, itself a result of East–West encounter. Professor Ray writes,

> Since the days of Sir W. Jones, Sanskrit literature, in almost every department, has been zealously ransacked by scholars, both European and Indian. As the results of their labours, we are now in possession of ample facts and data which enable us some idea of the knowledge of the Hindus of the old in the fields of philosophy and mathematics including astronomy, arithmetic, algebra, trigonometry and geometry. Even medicine has received some share of attention.[17]

But why did, after all, Professor Ray write the book on the history of Hindu chemistry? He was quite aware that the branch of chemistry was "entirely neglected" in the oriental studies, partly because of "its complex and technical nature".[18] He had a fascination for the understanding of the "progress of chemical knowledge among ancient nations".[19] There is reason to believe that Professor Ray was guided both by his nationalist feeling and his own evaluation of the colonial impact on India. He himself undertook a systematic examination, from the chemical standpoint, of the works of Charaka and Sushruta, and the various standard works of the Ayurvedic and Iatro-chemical Periods.

It was at this juncture that he came into contact with the famous French historian of chemistry, M. P. E. Berthelot. Professor Ray writes,

> The illustrious French savant, the *Doyen* of the chemical world who has done more than any other person to clear up the sources and trace the progress of chemical science in the West, expressed a strong desire to know all about the contributions of the Hindus and even went the length of making a personal appeal to me to help him with information on the subject. In response to his sacred call I submitted to him, in 1898, a short monograph on Indian alchemy: it was based chiefly on *Rasendrasara Samgraha*, a work which I have since then found to be of minor importance and not calculated to throw much light on the vexed question as

> to the origin of the Hindu chemistry. M. P. E. Berthelot not only did me the honour of reviewing at length but very kindly presented me with a complete set of his monumental work, in three volumes, on the chemistry of the Middle Ages dealing chiefly with the Arabic and Syrian contributions on the subject, the very existence of which I was not till then aware of. On perusing the contents of these works, I was filled with the ambition of supplementing them with one on Hindu chemistry.[20]

In the preface to the second edition of the work, Professor Ray went to the extent of throwing further light on the "independent origin of the Hindu system and its priority to that of the Greeks".[21]

Professor Ray was quite aware of a major criticism of the first edition of his work (Volume I) by no less than a person by the name M. P. E. Berthelot: it is the absence of a systemic study of scientific theories. In the second volume of his work, published in 1909, Professor Ray added extensive extracts from the writings of Professor Brajendranath Seal. These include the mechanical, physical and chemical theories of the ancient Hindus incorporating (a) the *Sankhya-Patanjala* system, (b) chemistry in the medical schools of ancient India, (c) the *Nyaya-Vaiseshika* chemical theory and (d) the conception of Molecular Motion. Professor Ray also added in the appendix a full chapter by B. Seal on "the Hindu doctrine of scientific method". In the preface to Volume II, Ray quoted preliminary remarks of Seal on the scientific method of the Hindus, "which shows that all this was not a mass of unverified and unverifiable speculation (the very antipodes of science) – the charge usually brought against Hindu thought and culture, but professed to be the outcome of scientific methodology".[22] Here, Seal also elaborated on historical methodology, "A true historical perspective can only be acquired by historic-comparative studies with the application of the correct historic-comparative method".[23]

It thus becomes apparent that the perspectives of East–West encounter and nationalist discourse, prevalent in the early 20th century, very considerably influenced the writings of Professor P. C. Ray on the history of chemistry in ancient India. He went so much as to write in the preface of Volume II that "the Hindu nation with its glorious past and vast latent potentialities may yet look forward to a still more glorious future".[24] This pride in the scientific achievements of ancient India did not lead him to be parochial as Volume II of his work was dedicated "to the memory of M. P. E. Berthelot". In that sense, any study of history of science in ancient India cannot altogether rule out the broader aspects of East–West encounter along with its colonial and nationalist dimensions as evident in modern India.

Notes

1 Helge Kragh, *Introduction to Historiography of Science*, Cambridge University Press, Cambridge, 1987, Chapter 3, pp. 32–40.

2 Quoted from G. Krobe's article entitled "*Wissenschaftswissenschaft und Wissenschaftsgeschichte*", cited in *Ibid.*, p. 36.
3 G. Sarton, *An Introduction to the History of Science*, 3 Vols., Williams and Wilkins Company, Baltimore, MD, 1927–1948, *passim.*
4 C. P. Snow, *The Two Cultures and a Second Look*, Cambridge University Press, Cambridge, London, 1966.
5 G. T. Clark, "The Science of History and the History of Science", in H. D. Roller (ed.), *Perspectives in the History of Science and Technology*, University of Oklahoma Press, Norman, OK, 1971, quoted in Helge Kragh, *op. cit.*, p. 37.
6 I. B. Cohen, "History of Science as an Academic Discipline", in A. C. Crombie (ed.), *Scientific Change*, Basic Books Inc., USA, 1963, quoted in *Ibid.*, p. 38.
7 Dhruv Raina, *Images and Contexts: The Historiography of Science and Modernity of India*, Oxford University Press (OUP), New Delhi, 2003, p. 2.
8 *Ibid.*
9 S. Irfan Habib and Dhruv Raina (eds.), *Social History of Science in Colonial India*, Oxford University Press (OUP), New Delhi, 2007, p. xiv.
10 *Ibid.*, p. xvi.
11 See, "*Introduction*" by Dhruv Raina in S. Irfan Habib and Dhruv Raina (eds.), *Situating the History of Science: Dialogue with Joseph Needham*, Oxford University Press (OUP), New Delhi, 1999, pp. 1–15.
12 Zaheer Baber, "Science, Technology and Colonial Power", in S. Irfan Habib and Dhruv Raina (eds.), *Social History of Science*, *op. cit.*, p. 151.
13 Romila Thapar, "History of Science and the Oikoumene", in S. Irfan Habib and Dhruv Raina (eds.), *Situating the History of Science*, *op. cit.*, p. 27.
14 "*Introduction*", in S. Irfan Habib and Dhruv Raina (eds.), *Social History of Science*, *op. cit.*, p. xxx; Romila Thapar, in S. Irfan Habib and Dhruv Raina, *Situating the History of Science*, *op. cit.*, p. 27.
15 Romila Thapar, in S. Irfan Habib and Dhruv Raina, *Situating the History of Science*, *ibid.*, p. 27.
16 Prafulla Chandra Ray, *A History of Hindu Chemistry*, Vol. I, The Bengal Chemical and Pharmaceuticals Works, Limited, Calcutta, Second Edition, 1903, p. 1.
17 *Ibid.*, p. 2.
18 *Ibid.*
19 *Ibid.*
20 *Ibid.*, pp. 4–5.
21 *Ibid.*, p. 6.
22 Prafulla Chandra Ray, *A History of Hindu Chemistry*, Vol. II, The Bengal Chemical and Pharmaceuticals Works, Limited, Calcutta, 1909, p. F.
23 *Ibid.*, p. G.
24 *Ibid.*, p. D.

3 Re-reading Iron in Northeast India

Beyond Nationalist and Communal Approaches

Amrendra Kumar Thakur

In this chapter, the subject of iron in Northeast India has been studied beyond the nationalist and communal approaches of history writing. It shifts its focus away from the attention that dynastic regional history received in the hands of the British historians, which was sadly the trend followed by the later-day nationalist and communal historians as well. This is a departure from the general oft-trodden path which includes the analysis of empires – their administration, social life, cultural contribution, invasions, etc.; the celebration of the careers and achievements of "great" rulers; the condemnation of the excesses of "bad" rulers and alternation of golden and dark ages. Not surprisingly, these historians have used the same sources repeatedly, analyzed the same, using similar parameters, and reached slightly different conclusions depending on their perspectives.[1] The successive generations of researchers, however, have adopted their own methods and consequently tried to provide a critique of these writings of the past. Of course, these new generations of historians have new sources, methods, tools and techniques at their disposal. As a result, there have been further contributions to our understanding of developments with respect to iron technology, but much more remains to be done.

The chapter is based on archival, library and field researches, and it comprises two parts. The first part briefly examines the factors related to the neglect of Northeast India in the studies on the technology of India and provides examples of the same. The second part gives a descriptive account of iron in the region. This discussion on iron in Northeast India is centred on the following four areas:

i. Sources and processing of iron
ii. Manufacturing of tools and weapons, and the groups or communities involved in the process
iii. Influence of neighbouring areas on iron technology of the region
iv. Impact of iron technology on the economy, polity and society of Northeast India

DOI: 10.4324/9781003370642-4

I

This chapter is aimed at examining the iron technology and its impact on Northeast India from a perspective that tries to go beyond the two above-mentioned approaches, also because they sometimes are unable to study the problem objectively, free from any bias and prejudices.[2] Moreover, the study of iron in Northeast India has not only remained outside the focus of the two above-mentioned yardsticks but also of the historians of other schools. Arun Bandhopadhyay, in his edited volume *Science and Society in India: 1750–2000*, for example, fails to provide a "regional balance" as pointed out by the reviewer. Interestingly, even the reviewer has found a few articles, particularly on Sir C. V. Raman and Ramanujan, to be sufficient for providing a "regional balance". It seems that the "Indian" northeastern region doesn't even figure in the schema of either of them. Similar observations can be made regarding Dilip K. Chakrabarti's book and its review.[3] Hence, it is high time that the historians of every school should come out of their cocoons and study the valleys and hills of Northeast India free from any bias and prejudice.

Studying the problem of technology, beyond confined approaches, is very important to perceive and analyze it comprehensively. It is so because no society grows in isolation. We certainly cannot agree on several views and approaches to the study of European technology as adopted by Lynn White Jr., but he is correct to opine that "technology knows neither chronological nor geographical frontiers".[4] The same is true in the context of this attempt to study iron technology in Northeast India due to its comparatively isolated geographical location, and a rather slow pace of technological development in the past, especially during the pre-colonial period.

It is not only the northeastern region of India that has always been neglected in the studies at all India level, but other tribal areas of India also faced the same academic apathy. It is clear from the outcome of the survey of research conducted by Ajit K. Danda on tribes in India. He has observed that the sense of aesthetics, crafts and arts, including fine arts, for which tribal people enjoy a distinction, have received practically no mention. He further adds that it was found that altogether 926 items dealt with the aspects of tribes in India during the decade under consideration; 336 among them can broadly be categorized as ethnographic accounts. The total number of articles highlighting aspects of tribal culture and change during the decade is 564, among which 167 deal primarily with change, and the remaining 397 reflect some aspect of culture or other. The other 26 items concentrated attention on problems as well as the development related to tribes. David R. Syiemlieh's survey of historical research on the land and people of Northeast India has also similar findings to suggest. It is clear that the historians of modern India, dealing with technology, could not pick up the trend of studying tribal iron technology from Verrier Elwin's

The Agaria, which remains one of the earliest works dealing with the same in depth.[5]

Generally speaking, technological inquiries in India have virtually remained confined to the macro levels. The studies have mostly been on the mainland of India and its dominant themes like neoliths, painted wares, iron, etc. The inquiries into the technologies of the peripheral and marginalized tribal areas are not very popular. To the European mind, the tribal people were nothing but queer and exotic. The necessity of colonial administration and economy nevertheless made the bureaucrats undertake intensive study of the life, culture and resources, including forest, mines and minerals of such people. However, to the colonial rulers, the alien subject population, particularly those under colonial subjugation, appeared only as savages. Lacking an awareness of heterogeneous realities, the very approach to the study of the "alien" tribes became an exercise in dealing with the cultural isolates. Even after independence, the researches on tribal technology and its social implications were never on the agenda of social science researchers on a continuous basis. This was because they had inherited not only concepts, techniques, theories and methods of study but also topics from the early ethnographers. Hence, the study of tribal technology did not receive the deserved attention of the scholars (belonging to any school), and the study of the iron technology of Northeast India is not an exception to it.

Some other reasons behind such neglects have been well explained by Eric Hobsbawm. He writes,

> So long as trade has not become worldwide and is not based on large-scale industry, the technological advances due to these developments remain insecure. They may, being locally or regionally based, be lost in the consequence of barbarian invasions or wars.[6]

The term "insecure", used by Hobsbawm, can be explained as unnoticed/un-researched in the case of Northeast India because the smelting, use and trade of iron never became part of the worldwide trade. However, it was thriving in Northeast India, including neighbouring areas of contemporary Bengal and Bangladesh during the pre-colonial period. And the expression "consequences of wars", in the case of Northeast India, can refer to a colonial onslaught. British iron easily replaced the indigenous iron, and the volume of import of iron in Assam, by the end of the 19th century, reached 2,85,527 mounds valued at Rs. 15,34,707.[7] The changed situation of the source of iron has certainly influenced the general lack of research interest in this field.

II

The term "Re-reading" has been chosen as one of the focused areas of this chapter because this is not the first attempt to study iron in Northeast India.

Before this, a number of British officers and Indian scholars studied the same. The British first encountered the people of Northeast India in 1767, following the establishment of the East India Company in Bengal. But it was only in the aftermath of the Treaty of Yandaboo (1826) and the resultant British occupation of various hills and valleys that the imperative of establishing their rule was felt by them. Thus, the British made a concerted effort to have a better grasp of the history and culture of the people in the larger context of the history of India's northeast region, then known as Assam. They started maintaining written documents in the form of reports, accounts, notes, diaries and books, and these were meant to serve as sources of information on the history, culture and, more importantly, the economic resources of the region.

Similar has been the case of iron with several officers writing about the practices related to it.[8] Such writings are important sources that lay the groundwork for further research work. Subsequently, after Independence, some Indian scholars working on the land and people of the region have studied the same. The problem, however, in these previous attempts to study iron in this region of India is their isolated nature. Scholars have mostly confined their works either to an area or to a particular tribe, scantily showing linkages between different areas of the region and influences of the neighbouring areas or tribes. The problem in the study of iron is that it has always been studied within the confines of contemporary political boundaries.

Meghalaya

Three shades of opinion have been expressed on the origin of iron smelting and iron use in Khasi-Jaintia society. Basing itself on linguistic grounds, the first view holds that iron has had an indigenous origin. Of the words for metals, those for iron (*nar*), brass (*rynnong*) and gold (*ksiar*) are indigenous, unlike copper (*tama*) and silver (*rupa*) which suggest Indo-Aryan origin. Though the term for iron may suggest indigenous origin, the myth, prevalent in the society, points to something else that forms the foundation for the second strand of opinion.

The ancestress of the Diengdoh ("wooden trough") clan of Cherra and the Lalu ("earthen pot") clan of Jaintia Hills were supposed to have come from beyond the Kapili (Kopili) River in Assam to the Jaintia Hills. There she married a Jaintia man. She is said to have had the power of reducing her size to such an extent that she could get into an earthen jar at will, and so she and her children got the nickname of Lalu. During a plague, all the members of the family died except for one female, called Ka Iaw-iaw, who inherited all the family property. Naturally, she had many suitors, but she fled from them to Jowai where she took refuge in the house of a *lyngdoh* (priest). Under pressure from his wife, who thought she was penniless, the *lyngdoh* tried to sell her as a slave. Nobody would offer more than 20 cowries for her, so the *lyngdoh* had to let her stay in his house. Out of gratitude for the shelter offered to her, she transferred all her wealth to his house. The *lyngdoh*'s son eventually married her. Soon the news reached the *lyngdoh* that some people from beyond the Kapili (Kopili) River were making preparations to kidnap his rich daughter-in-law. He arranged for her to escape to a place in the Khasi Hills near Nongkrem, known as Sohphohnongrum. There she found a business where each labourer was paid 20 cowries per day.

It is said that it was this enterprising woman who first introduced the art of iron smelting into the Khasi Hills on her estate at Sohphohnongrum. The third view traces the beginnings of iron to the larger metal tradition in Southeast Asia, associates it with the Dongson culture in the 4th or 3rd century BC and opines that probably iron was not widely used until around the beginning of the Christian era.[9]

The principal sites for the mining operations were around Mylliem, Nongkrem, Laitlyngkot and around Mairang and Cherrapunji. Open-cast mining was invariably worked in the monsoon, with the flow of rainwater down the hillsides assisting the extraction of iron ore. After washing the ore a number of times, the material would be smelted. Lt. Yule gives a detailed account of the process of smelting the ore. His account differs slightly from that of the account provided by William Cracroft, an early British administrator in the Khasi Hills. Cracroft's account describes the process that used two large upright bellows; the furnace made of pipe clay braced with iron hoops and the chimney. His account is as follows:

> At the right side of the bellows and even with the top of the chimney, is a trough containing damp charcoal and iron sand: at every motion of his body the operator with a long spoon tumbles a piece of this charcoal with the iron sand adhering to it, down the funnel of the furnace and when a mass of melted or rather softened iron is formed on the hearth, it is taken out with tongs and beaten with a heavy wooden mallet on a large stone by way of anvil.[10]

Women and young boys were also employed in the industry. Iron ore was often not smelted in the villages adjoining the mines. It was sold in baskets containing three maunds of ore and carried often for many miles to the villages where the smelting furnaces were located. Again, the manufacture of artefacts in the hills was done in workshops differing entirely from the huts in which the first smelting was done. The artefacts manufactured in the hills were swords, arrowheads, spears and tools.

Robert Lindsay, Collector of Sylhet in the 1770s, recollected in his memoir, *Anecdotes of an Indian Life*, that the Khasi Hills produced wood of various kinds adapted to boat and shipbuilding and also iron of very superior quality. With regard to iron, he further wrote,

> It is brought down from the hills in lumps of adhesive sand and being put into the forge, produces excellent malleable iron without ever

> undergoing the process of fusion, the hammer and fire discharging the dross and coarser particles at once, thus producing what is called virgin iron, superior to any made in Europe by charcoal.[11]

Iron constituted the principal industry in the Khasi Hills and was the chief export item. Limestone too was exported in large quantity though not from as early a date as that of iron. Apparently, the trade was in favour of the hill people. They appear to have exported more than what was imported. The items of export would have fetched more, even in barter terms, to enable the Khasis to convert some of the profits of the trade into gold and silver. This could be an explanation for why the Khasis–Jaintias have a fondness for jewellery.

P. R. T. Gurdon in his monograph, *The Khasis*, noted in the early years of the 20th century that the export of iron in any form from the district has now almost dried out, only a few hoes being brought down by the Khasis from Laitdom in Khadsawphra to the Burdwar and Palasbari markets in the Kamrup district of the Assam valley. The disadvantages of expanding and improving the Khasi iron technology were many. Mining was expensive and laborious, for it was only surface mining with an insufficient concentration of iron ore. Transportation added to the cost, and the impure state in which the iron was sent out added to the weight and ate on the profits. The decline of the Khasi iron production and trade became evident soon after the British took over the control of the hills. What really rang the death knell for this industry was that it had to compete with the superior and cheaper English iron which was soon finding markets in the plains areas where the Khasi iron had a monopoly in the past. By the second half of the 19th century, the production and trade in iron had all but ceased as the import of English iron had far exceeded the amount of iron that was being produced locally.

The Garos of Meghalaya used traditional iron implements in their everyday life. This community had traditional blacksmiths until the 18th century. Their important iron-made agricultural implements were *ate* (*dao* or chopper) made of iron blade, *gitchi* (an iron scraper), *attema* (a large *dao* used for cutting big trees) and *ma* (an axe made of solid iron for cutting big logs). With regard to weapons, the *mil'am* is a two-edged and one-piece sword made of metal from hilt to point. The role of Garo in the iron trade is well explained by S. F. Hannay along with the difference in the nature of iron and its uses. He reports,

> In lower Assam, the Garrow Haths of Pulashbarree and Gohain supplied iron hoes manufactured by the Gossyahs, and these were largely imported in to Upper Assam for Government purposes. This iron from its soft or malleable nature was considered the best for the manufacture of nails, fire-arms of small size and the inner tubes of the large cannon; the iron of the Upper Assam was found best adapted for swords, axes, knives, shovels, hoes, etc.[12]

Thus, Hannay's observations also explain the trade networks and the sophistications in the iron technology of the Garos.

Assam

The natural availability of iron mixed clay (*loa-mati*) and iron (*lo*) in Doiyang Dhansiri valley (northern bank of the river Brahmaputra in Upper Assam), and the existence of the process of smelting of iron at Bossapathar, Doiyang, Dhekial, Kamargaon, Kamarbandha, Hamdai and at some other places, supports the view about the growth of native production and trade in iron since the early Christian era in Assam.[13] Slag heaps, scattered over an area of about one-fourth of a square kilometre around Radha Pukhuri, indicate the vastness of this iron extraction and smelting site. Some slags are heavy in density, and some are found embedded in burnt clay. Further sophistication in the working of iron during the ancient period of Assam is also witnessed. However, the impact of neighbouring areas on the working of iron cannot be ignored.

A pair of brass-coated iron architect's plummets was found at Kharua gaon in Barpathar while digging a pit. The iron core of the plummet is exposed in its round socket. The plummets have pointed ends at the bottom, flanked by two pointed ends terminating in a lotus bud each. Their central parts are adorned with a number of horizontal lines. The pair of plummets in their shapes and forms is similar to a Gupta metal plummet found in the Surma River. On this basis, the pair found at Barpathar may be dated to AD 600. The pair of Barpathar plummets, being the direct imitation of the rare example of Gupta metalwork made of bronze-coated iron, confirms that the alloying tradition in metal like that of the Gupta possibly flourished in the valley. Besides this technological advancement, the pair of Barpathar plummets suggests the availability of iron, blacksmiths and the manufacturing process of metallurgical crafts, apart from the existence of a specialized class of builders of brick architecture in the valley during AD 600 who were conversant to the use of such instruments in raising architecture.[14]

The prevalence of iron in Assam is also supported by the mention of Assam iron in the texts of ancient and medieval periods.[15] According to Plinny, the iron of Serica (Assam) was considered to be the best. The articles of merchandise, mentioned by Ammianus Marcellinus, from Serica (Assam), consisted of skin, iron, aloe, musk and horns of the rhinoceros. These classical sources, therefore, point to the work on iron from ancient times. The writers of the medieval period were also attracted by the iron industry of Assam. The Ahoms were excellent casters, as specifically reflected in the great skill they displayed in crafting match blocks and barchadar artillery. Along with firearms, the manufacture of gunpowder was an equally important industry during the time of the Ahoms. Thus much before the British writers, other writers have written about the iron in Assam during the precolonial period.

The first Ahom ruler of Assam, Sukapha, promoted the smelting of iron and the manufacture of iron implements on a large scale. The hill where iron was smelted was called Tiru hill. In the Tai language, the meaning of *tiru* is the place where iron was smelted (*ti* means smelting of iron and *ru* means caves). There was a small hill stream flowing through this area and this stream was also called Tiru. This stream ultimately drained into the Dikhou River. Sukapha established villages of iron smelters, including iron smelting workshops in this area. Remains of some iron smelting furnaces can still be seen here. A big pond is also found here, which was possibly dug by some Ahom monarch as a source of water required for the process of smelting of iron. Due to the presence of many iron smelting furnaces and workshops, this place was known as Tiru loha khat (*loha khat*, the place where iron is manufactured).

The workers employed in the Tiru loha khat as iron smelters were known as *tiruwal.* Some historians opine that the tiruwals or the iron smelters belonged to the Bodo tribe of Assam. There were three big villages of tiruwals, namely Lohakhat, Pachikhat and Perakhat. The industry was practised in two parallel forms: iron smelting was run by a distinct class of artisans known as loh shalias, and the smithy work was done by the blacksmiths locally known as kamars. Usually, all the blacksmiths did not perform the welding of iron. There were special groups of blacksmiths to do the welding works known as *dhekar garha kamar*. The smelted iron was cast into various shapes and forms by adopting various methods. The format which gave shape and size to the object was earthen. Some of the cast manufactures of Assam are cannons, guns, axles of cartwheels and various domestic goods.

Royal officers were commissioned by the Ahom rulers to look after the iron workshops and the workers. The *Tiruwal Rajkhowa* was the overall administrative in charge of the entire iron workshop region, and the Boras and the Saikias were in charge of the iron workers' villages. The officer who maintained the accounts was designated as *Tirukakati*. The paiks or the workers, who were employed in the iron workshops, had to deposit a certain amount of iron to the king, and they utilized the remaining for their own purposes or sold/exchanged with others. Iron in huge quantities was required for the manufacture of a large number of firearms and other weapons by the Ahom state.

Hiteshwar Barbarua has provided an interesting list of the types of artillery guns and firearms used during the Ahom days. Five types of cannons are listed: Biyagom bartup, Hatimuria bartup, Turbuki bartup, Mithahuleng tup and Baghmore tup. Eleven types of hilois or muskets are also listed. They were *Gothia hiloi*, *Jonbur hiloi*, *Polongi hiloi*, *Ramsangi hiloi*, *Soru Hatnolia hiloi*, *Kamayan hiloi*, *Zomur hiloi*, *Kasai hiloi*, *Tuwa hiloi*, *Khoka hiloi* and *Bosadari hiloi*.[16]

Other than the above-mentioned weapons, swords (royal [*hengdan*], warrior's [*tarowal*], royal sword ornamented with gold [*sonia hengdan*], royal sword ornamented with silver [*rupar hengdan*], long bent sword [*chandrahans*], double edged sword [*bumukhia tarowal*] and very large sword [*dakhar*]), dagger (*churi*), chopper (*daa/dao*) – the most common weapon – (*nakoi daa, mit daa, kopi daa, mechi daa, khari daa, khangara daa, bali kata daa* [sacrificial *daa*]), axe (*kuthar*), knife (*katari*), spear (*jathi*), leaf-shaped spear (*barsha*), barbed spear (*xel*), simple rod spear (*xul*), trident spear (*trishul*), multiple barbed spear (*khapar*), arrow heads (*kar/teer*) and iron rod fencing (*xingari*); the Ahom forts were invariably protected by gorges and iron rod fencing; fishing implements dentate fishing rod (*pocha*), single fishing rod (*koch*), fishing harpoon (*obhota kol*), fishing hook (*barasi*), iron weights of the fishing net (*jalar guli*), etc. also consumed a large amount of iron in their making. Not just the state but the commoners too used iron in everyday life.

As far as the use of iron as household implements is concerned, the following are important to be mentioned: *kor* (spade), *phal* (ploughshare), *chiprang* (pick axe), *joboka* (hoe), *kanchi* (sickles), *gul* (pounder head), *akuki* (anchor) and *sikali* (iron chain). Iron was also used as accessories for horses and elephants in pre-colonial Assam, such as *khura* (horse shoe), *hatir tongi* or *haoda* (elephant seat), *bagh jari* or *lekam* (reins), *langching* (elephant pincher), *hati bandha sikali* (elephant chain), *rekabi* (saddle), etc. Excellence in cannon making was witnessed with the expansion of the Ahom kingdom. At Tirugaon and Hattigar, traditional iron making was still being practised "for classing the ironworkers under Hazaree Keihs and Saikeahs and they amounted at one time to 3,000".[17]

Initiatives of the Ahom rulers as well as the contribution of certain specialized communities made the iron industry in Assam reach its pinnacle during the pre-colonial period. It was in the early 16th century, with the subjugation of the Chutias, that the kamars were first settled in the valley as a distinct *khel*. The *Chutia* prisoners of war, excellent iron workers, were settled in Bossa and some other places in Upper Assam. They were liable to supply to the royal storehouse swords, guns and cannons from the iron smelted by the loh shalias who resided in their vicinity.[18] This was the starting point of the formation of the *kamar khel* in the Ahom kingdom. During the reign of the Ahom king, Pratap Singha, a number of artisans were brought from the neighbouring states (Koch for instance), and people from the *kalita* caste (a higher in the social order) were taught blacksmithy.[19] Buchanon Hamilton has also observed that blacksmiths in Assam were mostly kalitas and Koch (Assam-Bengal border) who could make locks, padlocks, sacrificial knives, matchlocks, spears, spite nails and clamps for building boats.

According to Buchanan Hamilton, there was an iron mine in Boyang which supplied ore in abundance to the whole of Assam. Though the decline of the Ahoms resulted in the loss of magnitude of the iron industry in pre-colonial Assam, the trait did not die. In the 19th century, writers like S. F. Hannay and William Robinson, who wrote on the subject, found the remains of furnaces in the vicinity of Jaypur, Borhat and along the bank of Suffrai River in Upper Assam which convinced them about the existence of this industry in the region. Besides the remains of the furnaces, the relics of locally manufactured iron guns found in these places, particularly in Sibsagar, also help us to come to the conclusion that Upper Assam was the home of this particular branch.[20]

One can easily conclude from the above discussion that iron constituted an important component of trade. It also contributed to the expansion of the Ahom state as firearms and other weapons were made of iron. This increased the production and enhanced the tax of the state. Contrary to this, though Sudeshna Purkayastha provides us with valuable information collected from the various published sources of the medieval and the British periods, the generalizations regarding the use of iron in the ploughshare and its consequences are not tenable. She writes, "Where plough cultivation was in vogue, neither heavy plough drawn by several bullocks nor seed-drills were in use in the region". Contrary to her position, we have the evidence to show that in the case of Arunachal Pradesh, especially among the Singhphos, Khamptis and Adis (the closest neighbours of the people of Upper Assam), iron was used as the ploughshare, and it was drawn by trained elephants, especially in the case of the Singphos and Khamptis. Ploughs drawn by elephants were popular among the two, whereas buffalo or ox/bull was used in case of the Adis. On the other hand, Purkayastha does not take into her consideration the ecological factors; the soil type, early and long rainy season, availability of wood, root, etc. have not been examined in her analysis.[21]

Arunachal Pradesh

A myth prevalent among the people in the Shimong area clearly indicates that Assam was the source of iron for them. As the myth goes, Peddong-Kebo was a Wiyu in the form of a yak. As he was grazing in the forest, he trod on an ant, and as a result, his neck became small, his head large and hind-quarters enormous. Shortly afterwards, Kojam-Koja met and killed him. His bones went to Assam and turned into iron.[22] The myth related to the body parts of the yak is not only relevant for the source of iron but also for the minerals like salt, etc. from neighbouring Assam and Tibet.

Among the Adi, Apatani and Nishing tribes of Arunachal Pradesh, the indigenous process used for making metal-alloy objects like ornaments, weapons, etc. was "Cile perdue", i.e. "lost wax process". There is no information among these people about the extraction of metallic ores, but there is enough information about the art of making alloys and fine metalworks. Many of these objects are still preserved, and some of these are believed to possess divine powers. Possession of these valuable items is the symbol of one's prestige in the village society. Historically and mythologically, the manufacturing time is unknown, but the colour and shape are still fine and the ornaments never get rusted.

The ornaments like *Tashang/Tadok/Teshi* (necklace), *Kobyang/Koj/Koji/Kope/Kopung* (bangles), objects like *Maji/Maj/Ery/Ripa/Emul* (bronze bell),

Talo/Tal/Tale (plate), and the weapons like *Elyo/Olyo/ Yoksha/Oyok* (*dao*), *Morto/Yongmo* (arrow-head), etc. possessed by the Apatanis, Nishings and Adis of Arunachal Pradesh are the finest examples of the rich heritage of craftsmanship of the Lomas of ancient times. The Loma/Lomang, i.e. "gods of construction work", were expert and skilled craftsmen who manufactured these things and commanded high respect or divine status in the society. In the Adi society, the craftsmen are called Yangmo/Rangmo. The Padam Minyong groups of the Adis believe that Ninur Lomang was the elder brother of the Abotani – their ancestor.[23]

This myth and the availability of forging in metal, especially in iron in central Arunachal Pradesh, clearly indicate that the major tribes, such as the Galo, Adi, Nyishi and Apatani, largely obtained iron from Assam and fashioned it in the hills. Unlike the central part of Arunachal Pradesh, the iron ore was found in the eastern part, and the Khamptis and the Singphos not only smelted the ore but also produced agricultural tools, implements and weapons. In the western part, the iron ore was under the control of the Tawang monastery, and it proved as an important tool in the expansion of the monastic influence.[24]

The control over iron by the Khamptis and Singphos in their hills, the technology evolved by them for the daos (hatchet) and other war weapons and the settled agriculture in the Dihing river valleys must have positively contributed to the socio-political formations and expansion of the pre-colonial states of eastern Arunachal Pradesh by the medieval period.[25] The permanent place of Khampti chief in the "Great Council" of Ahoms is also indicative of the power and privileges enjoyed by them. The works in the iron of the Khamptis, during the pre-colonial period, attracted the British after the occupation of Assam. Even the British were proposing to procure the khampti daos for the British soldiers stationed in the area.[26]

The proposal of 1840 clearly mentions,

> I would beg to recommend that Khampti Daos be purchased for the whole of the sepoys at an estimated expense of 1,350 rupees as proposed by Captain Vetch ... that the Daos should always be kept up complete hereafter and be mustered as a part of the regular equipment of the Regiment.

The description of Verrier Elwin, regarding other aspects of the Khampti technology, is also very important. He writes, "Wood carving and ivory-work have persisted to this day ... . The Khamptis still make embossed shields and are fond of masks, mainly of the horror type". He also mentions specific practices of the Khampti chiefs,

> It is customary for the chiefs also to employ themselves in useful and ornamental arts. They work in gold, silver and iron, forge their own weapons and make their wives' jewels. They also manufacture embossed

shields of buffalo or rhinoceros hide gilding and lacquering them with skill and taste.[27]

Mizoram

Every Mizo village used to have an official village blacksmith (*Thirdeng*), although there could be one or two private Thirdengs who performed their service on a part-time basis. The service and skill of the official village *Thirdeng* could be availed by anyone in the village, and in exchange for his service, the *Thirdeng* got a basket of paddy from each household per annum. He was also entitled to receive a small share of every animal shot or trapped by any villager. The nature of his work was mainly to manufacture and repair agricultural implements, guns and other weapons. The iron was procured from Burma, and the blacksmiths were found throughout the hills, but Wunla, in the southern Chin Hills, was the only village that has made its name famous for ironwork.[28]

Blacksmiths mainly manufactured tools for household use and agricultural implements, such as *chempui* (*dao*), *hreipui* (axe), *hachhek* (adze), *tuthlawh* (hoe), *fawvah* (sickles), etc. Weaponry, such as *kingkawthnam* (long blade or knife), *silai zen* (gunpowder), arrow, spear, were also manufactured by them. Some of the blacksmiths were also able to manufacture gun locks. Very good moulding in brass was occasionally to be minted with, and some of the smiths were able to manufacture gun locks. Among the Mara tribes, the articles which the blacksmith (Seudaipa in Mara) manufactures are daos, knives, hoes and axes. The tools used by the blacksmith are a hammer, pair of tongs, a file and a kind of chisel. His assistant works the bellows, pulling the pistons up and down, and the blacksmith holds the iron with the tongs in his left hand and a hammer in his right hand.

Usually, a separate thatched house called *pum* was made in the middle of the village where access was easy from different corners. It was the chosen spot for the village blacksmith workshop. It also functioned as a kind of resting place and chat house for old people. The conversations in the *pum* were also sources of information for the villagers. The blacksmith's forge consisted of two bamboo bellows made out of two large bamboos standing upright with pistons made of bamboo and cock's feather and worked by hand. Charcoal was used for the furnace, and the anvil consisted of a block of stone or a log of wood covered with a sheet of iron.

The blacksmiths played an important role in the evolution of the chieftainship and its development. They were important officials in the village. The blacksmith was appointed by the chief, and his job was usually hereditary. In some cases, one could learn the skill and work as a blacksmith in the village. The number of blacksmiths depended on the size of the village, but it was usual to have one or two in a village. In a big village, the chief had his own blacksmith known as "*lal thirdeng*" where the tools of chiefs and his elders were repaired and manufactured. Particularly, the *lal thirdeng* in a big village was closer to the chief compared to the commoners. If the blacksmith was greatly skilled, the chief was very proud of him and was afraid of his migration to other villages. Being an official in the village, he was one of the members of the council of the chief to participate in the meeting.

Moreover, one has found that many of the stone monuments of the Mizo, located in the present Chin Hills of Burma, bear engravings of human figures, animals, tools and weapons like spear, *dao*, etc. Subsequently, some of the carvings on the stones of the Mizo located in the Chin Hills were also made in relief or embossed. To make carving or relief on stone, one must be equipped with some kind of tools like a chisel or tools made of iron. This indicates that the Mizo were already familiar with the use of iron tools and weapons while living in Chin Hills of Burma, prior to their migration to the present Mizoram, and those tools seemed to have been made of iron. Apparently, if they used iron for weapons as well as for agricultural tools, there should be a skilled person who repairs and manufactures such tools. Therefore, we can say that blacksmithy was a popular trait of Mizo society during the pre-colonial period and even after it.

Nagaland

Metallurgy in Nagaland has been discovered from the usual Neolithic and Chalcolithic stages to the Iron Age in the southwestern part of Nagaland. The smiths produced daos, axes, sickles, knives, spears and so on. Iron implements used by the Nagas were spearheads, daos (*zhe*), *merr* (axes), *sarokejii* (spade-hoes), *tsukwe* (knives), *riizhe* (chisels), *chtigeh* (drill) points and awls.[29] These objects were used for agriculture and defence. Blacksmithy was a very important activity, particularly for the Angamis. Tradition has it that the Rengmas were considered the best Naga blacksmiths who smelted iron from the iron-bearing sands. The western *Rengma* and Ao Nagas used a specially made large spearhead of iron as currency. The Ao Nagas also used iron coins called *jabili*. Two such iron coins are preserved in the Assam State Museum, Guwahati. The eastern Rengma Nagas, on the other hand, used daos as currency.[30] In the recent excavations at Chungliyimty in Nagaland, a *jabili* associated with kaolin ware was found. This find has been dated between the 1st and 3rd centuries.

The Ao Nagas were not conversant with metalworking before the 18th century. It is known that four strangers from the plains brought iron technology that the Aos accepted. That technology further helped in manufacturing the daos and a few single piston bellows. There are two or more blacksmiths in each Angami village who cater to the basic requirements of the inhabitants. Hutton has referred to the technology practised by the Angami Nagas. Perhaps, they did not know the iron smelting technique. They made the heating furnace with flat sandstone with holes set upright against which charcoal was heaped. They used to get iron from Manipur or Assam. The Lhota Nagas were also conversant with daos (*lepok*).

The forge shops of Naga smiths were equipped with *rekri-chi* (stone anvils), *rekri jivii* (stone hammers) and *jibbeh* (pincers) made of cane and *kuru* (bellow) made of bamboo. However, the Angami smiths possessed the knowledge of quenching and tempering. They used salt, chillies and bamboo pickles as the quenching medium. Grinding was done on the stone to sharpen the tools. Spearheads were sharpened on both sides, whereas the daos were sharpened on one side. After forging and shaping, the cold sickles were serrated on one side with the help of *dao*.

Chingmei, which is apparently on the watershed dividing the sources of the Tsuta (Tita) and the Zungki from another stream which runs directly east to the Chindwin, is in touch with the plains of Burma side. Mills picked up a *Khamti dao* here, and the metal armlets they wear hereabouts are said to be from a place called Kamlugh which is believed to be Khamtilong. Hutton saw a *Khamti dao* again here, bought by a Chingmei man from Noklok further east.[31] The team members of Nagaland University have initiated a study of these objects. There is yet no firm radiometric data to trace the beginning of iron in the Naga Hills. According to Jamir, oral sources inform that in ancient times metals were brought to Nagaland from Manipur and Assam. For the Naga communities, being close to Myanmar, the probable accessible route for metals would have been through there.[32]

Manipur

At Sekta in Manipur (24°53′24″N, 94°24′E), iron spearheads and knives were excavated from the burials of the later period. At Thoubal, in the Khangabok district, the excavation revealed a wide variety of iron blades.[33] In this state, pre-industrial metal crafts were popular until the 19th century. During iron smelting, they used limestone as a flux. Principal pre-industrial iron-made products of Manipur were axes, hoes and ploughshares. Even today, iron objects are quite popular in this state. Some of these are the spear (*rung* or *bui*), harpoon (*to*), etc. However, *dao* is the most important iron tool that every Manipuri carries with him. The main iron implements used by the Meiteis are *langol* (plough), *thanggol* (sickle), *thangchao* (large *dao*) and so on.[34]

The discovery of Neolithic tools from almost the whole of Northeast India, bearing Southeast Asiatic tradition, justifies prehistoric human migration from Southeast Asiatic countries to this region of India since ancient times. When the Chinese established Yung Ch'ang in AD 68, for some centuries after, the population of the hilly areas was probably very small. It would have tended to concentrate into the areas of greatest economic importance, e.g. near trade routes, salt wells and iron ore deposits. This interactive influence is becoming clearer as the vast extent of pre-colonial trade and migration of people is being properly understood by researchers. The "proto-Jinghpaws" were a group who exploited the iron ore deposits in the vicinity of Hkamti Long.[35] The role of trade in Northern Chin society cannot be fully gauged without making reference to the external trade with Burma and India. The most important item traditionally imported from Burma is iron in the form of raw steel ingots, knives and other tools, which Chin smiths forge or re-forge into steel tools for war, farming and general use. F. K. Lehman is of the view that the social, economic and political development of Northern Chin lineages resulted from their desire to further economic intercourse with Burman civilization and also with Assam, Manipur, Arunachal Pradesh and neighbouring areas.[36]

Perhaps, the elaboration of the Northern Chin technology and institutional facilities for expanding and preserving the cultural inventory came from the need to make up in local technical and social resources for the uncertainty of the flow of goods from regular markets in the plains of Burma. Many villages have blacksmiths who are hereditary village servants, and they get paid for their repair of arms and agricultural implements by a tax of grain and meat levied by the headman on the villagers. This work is done mostly between harvest and the next planting, but in seasons when agricultural work is light, they also take private orders at piecework rates. This smith is a kind of village officer, apparently common in Falam, but no longer common in the Haka area. And we see the development of iron technology on similar lines in various areas of Northeast India.

Similar techno-cultural influence can be observed in the areas of upper Burma. Jinghpaw-speaking Kachins, inhabiting an area east of the Mali Hka (Irrawaddy) and north of the Shang Hka (Nam Tisang), are the closest neighbours to the Shans of Hkamti Long and have a widespread reputation as blacksmiths. They are known as the *Singhpo* in Arunachal Pradesh. There are various myths related to the supernatural powers of the blacksmiths prevalent in this part. It is believed that the blacksmith emerged as a revolutionary leader in favour of the slaves (*Gumsa – Gumlao*). The point to note here is that Dumsa La Lawn, the revolutionary leader, was a priest. This individual seems to be identified by Kawlu Ma Nawng with N'Dup Nawng Dai Gawng, one of the originators of the N'Dup-Dumsa ("blacksmith-priest") lineage, the originators of the Gumlao movement. This makes it clear that blacksmiths have played an important role not only in social formation but also in social mobilization. The influential social position of the blacksmiths is witnessed in a number of societies of Northeast India.

Besides the influence of the Southeast Asian traditions, the techno-cultural traditions of greater parts of India also had an impact on the technological progress of Northeast India in general and iron in particular during early times. During the reign of the Ahom king, Pratap Singha, a number of artisans were brought from the neighbouring states and people from the *kalita* caste (a high in the social order) were taught the blacksmith's trait. The blacksmiths of the Koch community of Bengal were also settled in Assam to work in iron, and they also influenced the craft of the blacksmiths. Similar are the cases of the Pangals of Manipur and some other Muslim artisans in Assam and other areas of Northeast India. The elaboration of the contributions of these various communities to the technological progress of Northeast India is deliberately left undone due to limitations of space.

Thus, to conclude, we can say that the "Re-reading" of iron in Northeast India is beset with the problem of diffusion from one culture zone to the other. The technological matrix crisscrossed several culture zones, and the question of its historicity continues to baffle historians, anthropologists and other social scientists. We can also say that besides hereditary/family or individual specialization in ironworking, the local specialization for the smelting of iron to smithy also developed. It certainly served as a sort of division of labour, allowing a rational allocation of productive effort in a few parts of the region. And, in Northeast India, there developed specialized skills in producers and consequent networks of trade during the pre-colonial period in which a number of villages and areas formed an interdependent economic network. The network worked successfully not only in the areas of what is now known as Northeast India but also in the areas of contemporary West Bengal, Bangladesh and beyond until the colonial state and their goods replaced the indigenous ones by the 19th century.

Notes

1 Harbans Mukhia, "The Teaching of History in Indian Universities: How Much of It Has Changed?", *The Indian Journal of Social Science*, Vol. I, No. 1, 1988, pp. 57–63; for other details, see V. C. P. Chaudhary, *Secularism versus Communalism*, Navdhara Samiti, Patna, 1977; Kesavan Veluthat, "R.S.Sharma – A Tribute", *Social Scientist*, Vol. 39, Nos. 9–10, 2011, pp. 83–86; Ashok Kumar, *Jinhone Likha Vanchiton ka Itihas, Hans* (Hindi), Akshar Prakashan, New Delhi, 2011, pp. 16–17.

2 The monograph *Science and Technology in Medieval India: A Bibliography of Source Materials in Sanskrit, Arabic and Persian*, by A. Rahman with the research assistance of M. A. Alvi (Persian), S. A. Khan Ghori (Arabic) and K. V. Samba Murthy (Sanskrit), published by the Indian National Science Academy, New Delhi, 1982, is the finest example of the nationalist approach to the study of science and technology in India. Susobhan Sarkar also expresses the same sentiments in the article, *Mysticism in the History of Science, Bengal Renaissance and Other Essays*, People's Publishing House, New Delhi, 1970, pp. 245–257. Whereas, D. P. Agrawal's, *Harappan Technology and Its Legacy*, Rupa, New Delhi, 2009, in association with Infinity Foundation, Princeton, NJ, and Vibha Tripathy's *History of Iron Technology in India (From Beginning to Pre-modern Times)* with a foreword by Dilip K. Chakrabarti, Rupa & Co., New Delhi, 2008, in association with Infinity Foundation, Princeton, NJ, are the best representatives of the combination of the nationalist and communal approaches. The details of the same have been presented in a chapter by A. K. Thakur, "Studying Technology of Hill Societies of Northeast India: Problems and Perspectives", in *Proceedings of North East India History Association*, 32nd session, Tripura University, Agartala, 2011, pp. 38–46. Hence, to avoid repetition, narrating the same has been dispensed within this chapter. Unlike these two above cited books, in the series on *Contributions to History of Indian Science and Technology*, which have numerous nationalist and communal biases marked by the neglect of Northeast India (Tripathi on p. 181 even provides wrong

information about Assam without cross-checking it). The book by Pranab K. Chattopadhyay and Gautam Sengupta, *History of Metals in Eastern India and Bangladesh*, Pentagon Press, New Delhi, 2011, in association with Infinity Foundation, Princeton, USA, provides information on iron in northeast India (pp. 58–63) and is largely free from the above-mentioned overtones. This paper draws largely from the book while discussing aspects related to Nagaland and Manipur.

3 For details, see various articles and the editorial in B. P. Sahu (ed.), *Iron and Social Change in Early India*, Oxford University Press (OUP), New Delhi, 2006; Arun Bandyopadhyay (ed.), *Science and Society in India, 1750–2000*, Manohar, New Delhi, 2010, and its review by N. Rajendran in *The Indian Historical Review*, Vol. 38, No. I, 2011, pp. 171–174; Dilip K. Chakrabarti's, *The Early Use of Iron in India*, Oxford University Press (OUP), Delhi, 1992, and its review by Suraj Bhan, *The Indian Historical Review*, Vol. XI, 1994, pp. 103–107; Rajendran writes that "at the same time a pronounced gap exists in constructing the epistemology of science in India. Few more articles – particularly on Sir C. V. Raman and Ramanujan, could have been invited to be included to provide a regional balance as well as areas such as environmental history and colonial forest management could have made this volume more comprehensive".

4 For details, see Ishrat Alam, "Technology and Economic Progress in Medieval India", in Prabhat Patniak (ed.), *Excursus in History: Essays on Some Ideas of Irfan Habib*, Tulika Books, New Delhi, 2011, pp. 196–206.

5 For details, see Ajit K. Danda, *Tribal Ethnography*, ICSSR, New Delhi, 1996; David R. Syiemlieh, *Survey of Research in History of North-East India 1970–1990*, Regency, New Delhi, 2000; Verrier Elwin, *The Agaria*, Oxford University Press (OUP)/Humphrey Milford, London, 1942. For some other details on the reasons for the neglect of tribal technology and potentials of research, see A. K. Thakur, "Indigenous Paper in Pre-Colonial Arunachal: A Study on Manufacturing, Trade and Other Usage", in *Proceedings of the North East India History Association* (hereafter, *PNEIHA*), Shillong session (Silver Jubilee session), 2004, pp. 255–264, and idem, "Technological Progress in Pre-Colonial Arunachal Pradesh: An Overview", in *Proceedings of Indian History Congress*, 66th Session, Shantiniketan, 2006, pp. 797–808.

6 Eric Hobsbawm, *How to Change the World: Tales of Marx and Marxism*, Little Brown, London, 2011, p. 145.

7 Report on *River Bourne Trade of Assam, Year Ending 31 March, 1896*, p. 7 quoted in Priyam Goswami's *Indigenous Industries of Assam: Retrospect and Prospect*, published for Maulana Abul Kalam Azad Institute of Asian Studies, Kolkata by Anshah Publishing House, Delhi, 2005, p. 60. Impact on Khasi iron, in particular, has been studied by D. R. Syiemlieh, "Khasi Iron Culture and Iron Trade with Sylhet in the Late Eighteenth and Early Nineteenth Centuries", in *PNEIHA*, 8th Session, Kohima, 1987, pp. 242–250.

8 William Robinson, *A Descriptive Account of Assam*, London, 1841, pp. 35, 412 (Reprint: Sanskaran Prakashak, Delhi, 1975); W. W. Hunter, *A Statistical Account of Assam*, Trubner & Co., London, 1879 (Reprint: B. R. Publishing Corporation, Delhi, 1975); see also Alexander Mackenzie, *History of the Relations of Government with the Hill Tribes of the North-East Frontier of Bengal*, London, 1884 (Reprint: Mittal Publications, under the title *The North-East Frontier of India*, Delhi, 1979); also see E. A. Gait, *A History of Assam*, Thacker Spink & Co., London, 1905; 2nd edition, London, 1926 (7th Reprint of the 2nd edition, Lawyers' Book Stall, Guwahati, 1997).

9 For details, see N. C. Shadap-Sen, *The Origin and Early History of the Khasi-Synteng People*, Firma KLM Pvt. Ltd., Calcutta, 1981, pp. 58–59 and 195–196. Further details on Khasi-Jaintia iron can also be seen in D. R. Syiemlieh, *Khasi Iron*

Culture and Iron Trade with Sylhet in the Late 18th and Early 19th Centuries, *op. cit.* The first comprehensive work on the Khasi was an ethnographic monograph authored by P. R. T. Gurdon. He gathered his data from first-hand observations in the field, and he also consulted the notes and accounts left behind by the previous officials. In respect of iron technology, he simply drew on the earlier writings, e.g., Colonel Lister's estimation of Khasi iron exported to the plains, the reports of Yule and Cracroft on the process of iron-smelting and Thomas Oldham's description of the quality of Khasi iron; P. R. T. Gurdon, *The Khasis*, London, 1906 (1st pub.), Reprint: Akansha Publishing House, New Delhi, 2010.

10 W. Cracroft, "Smelting of Iron in the Kasya Hills", *Journal of the Asiatic Society of Bengal* (Hereafter, *JASB*), Vol. I, 1832, pp. 150–151. Another colonial account regarding the Khasi iron is also important enough to be mentioned: Lt. Yule, "Notes on the Iron of Kasia Hills", *JASB*, Vol. XI, Pt. ii, No. 129, 1842, pp. 853–857.

11 For a comprehensive history, see the notes of the first British explorers, the Lindsays, dating back to the period as early as 1778, stand as testimony to the superior quality of iron located in the "mountains" of the "Khasi"; L. Lindsay, *Lives of Lindsay or a Memoir of the House of Crawford and Balcarres*, Vol. III, John Murray, London, 1849.

12 M. S. Sangma, "Crafts and Technologies of Garos", in J. P. Singh and G. Sengupta (eds.), *Archaeology of North Eastern India*, Har Anand, New Delhi, 1991, pp. 304–312; also P. K. Chattopadhyay and G. Sengupta, *History of Metals in Eastern India and Bangladesh*, *op. cit.*; S. F. Hannay, "Notes on Iron Ore Statistics and Economic Geology of Upper Assam", *Journal of the Asiatic Society of Bengal*, No. 4, 1856, p. 333.

13 H. N. Dutta, *History, Art and Archaeology of Doiyang – Dhansiri Valley, Assam*, LBS Pub, 2012, pp. 154–155.

14 *Ibid.*

15 For details, see Dinesh Baishya, *Traditional Science and Material Culture of Early Assam*, EBH Publishers, Guwahati, 2009, pp. 299–335, the chapter on *Traditional Metallurgy and Metal-Wares*; Priyam Goswami, *Indigenous Industries of Assam: Retrospect and Prospect*, *op. cit.* To the surprise of the researcher, D. Baishya's book does not mention a comprehensive report on iron in Assam by S. F. Hannay. Nonetheless, D. Baishya presents an exhaustive account of iron along with other metals. Another great surprise is the opinion of Priyam Goswami that no iron was smelted in Assam (pp. 58–59); Sudeshna Purkayastha, "Iron Industry in the Brahmaputra Valley: Lon and the Kamars (Late Medieval to Mid Colonial Period)", *The Indian Historical Review*, Vol. *Shalias*. XXXIV, No. 1, January 2007, pp. 152–166. The discussion on Assam is largely from these works, S. Kakoty's work and accounts of the British period.

16 Hiteshwar Barbarua, *Ahomor Din*, Guwahati, 1997, p. 417, quoted in Sanjeeb Kakoty, *Technology Production and Social Formation in the Evolution of the Ahom State*, Regency Publication, New Delhi, 2003, pp. 116–117. The book by S. Kakoty provides an excellent description on the technological progress during the Ahoms and its contribution to the expansion of the Ahom state.

17 S. F. Hannay, *op. cit.*, pp. 330–344.

18 Hiteshwar Barbarua, *Ahomar Din,* Publication Board Assam, Guwahati, 1997 (Reprint, first published in 1949), p. 466; Benudhar Sharma, *Maniram Dewan*, Manuh Prakashan, Guwahati, 1950, pp. 275–276.

19 E. H. Darrah, *Notes on Some Industries of Assam*, Shillong, 1894.

20 S. F. Hannay, *op. cit.*, pp. 330–344; For details, see Buchanan Hamilton, *An Account of Assam*, Department of Historical and Antiquarian Studies, Guwahati, 1963; William Robinson, *A Descriptive Account of Assam*, London, 1841, Reprint, Sanskaran Prakashak, New Delhi, 1975.

21 For details, in case of Assam, see Sanjeeb Kakoty, *op. cit.*; for Arunachal Pradesh, see A. K. Thakur, "Ecology, Technology and Societies of Arunachal Pradesh", in *PNEIHA*, 31st Session, Tura, published, Shillong, 2011, pp. 103–114; *idem*, "Agrarian Technology in Arunachal Pradesh", *Arunachal University Research Journal*, Vol. 5, No. 1, 2002, pp. 129–134; "Process and Agency of Pre-Colonial States in Arunachal Pradesh", *The NEHU Journal*, Vol. I, No. 1, 2003 (new series), pp. 1–17; *idem*, "Technological Progress in Pre-Colonial Arunachal Pradesh: An Overview", in *Proceedings of Indian History Congress*, 66th Session, Shantiniketan, 2006, pp. 797–808.

22 Verrier Elwin, *Myths of the North-East Frontier of India*, Directorate of Research, Govt. of Arunachal Pradesh, Itanagar, 1993, Reprint, p. 240 (First published, 1958).

23 Y. P. Kohli, *Traditional Metal Alloy Craft of Arunachal Pradesh* and a number of articles on the traditional knowledge system of Y. P. Kohli and others have been published in various issues of the *Arunachal News*.

24 For details, see A. K. Thakur, *Technological Progress in Pre-Colonial Arunachal Pradesh: An Overview*, *op. cit.*, and Lt. E. T. Dalton wrote, "On the Meris and Abors of Assam", *JASB*, Vol. XIV, 1845 (New series).

25 For details, see Muhammad Kazim, *Alamgir Nama* (eds.) Khadim Hussain and Abdul Hai, Calcutta, 1865–73, pp. 723–783, quoted in Shireen Moosvi's *Man and Nature in Mughal Era*, Thematic symposium, Indian History Congress, 54th Session, Delhi, 1993.

26 *Foreign Dept. Political Br.* 1 March, 1840, No. 169 (National Archives of India, New Delhi).

27 Verrier Elwin, *The Art of the North East Frontier of India*, *op. cit.*, p. 84.

28 For details, on iron in Mizoram, see Bertram S. Carey and H. N. Tuck, *The Chin Hills*, Vol. I and II, Firma KLM (On behalf of Tribal Research Institute, Aizawl), Calcutta, 2008, Reprint (First published, 1932) and Malsawmliana, "Thirdeng (Blacksmith) in Traditional Mizo Society", in Malsawmliana and Benjamin Ralte (eds.), *Social, Economic and Political History of the Mizo*, EBH Pub. (India), Guwahati, 2011, pp. 11–17. We have been benefitted largely from these works.

29 For details, see P. K. Chattopadhyay and G. Sengupta, *History of Metals*, *op. cit.*, pp. 58–63.

30 J. P. Mills quoted in *ibid.*

31 J. H. Hutton, *Diaries of Two Tours in the Un-administered Area East of the Naga Hills*, Mittal Publications, New Delhi, 1995, pp. 20, 56.

32 For details, see T. Amir, "A Burial Site at Jotsoma and the Mortuary Customs of the Angami Nagas", in G. Sengupta, S. Roychoudhury and S. Som (eds.), *Past and Present Ethnoarchaeology in India*, Pragati Publications, Delhi, 2006, pp. 449–463.

33 *Indian Archaeology – A Review*, 1994–1995, pp. 58–59, and 1995–1996, p. 59, quoted in P. K. Chattopadhyay and G. Sengupta, *History of Metals*, *op. cit.*, pp. 58–63.

34 J. Mao, "Traditional Craft and Technology in Manipur", in J. P. Singh and G. Sengupta (eds.), *Archaeology of North Eastern India*, *op. cit.*, pp. 304–312.

35 E. R. Leach, *Political Systems of Highland Burma: A Study of Kachin Social Structure*, University of London, The Athlone Press, London, 1970.

36 F. K. Lehman, *The Structure of Chin Society*, Firma KLM Private Ltd, on behalf of Tribal Research Institute Aizawl-Mizoram, 1980 (First published, 1963) and some other similar details in H. N. C. Stevenson, *The Economics of the Central Chin Tribes*, Tribal Research Institute, Directorate of Education. Govt. of Mizoram, Aizawl, 1986.

4 Iron Technology and Urbanization in Odisha

(1st Millennium BC to 4th Century AD)

Byomakesh Tripathy and Dibishada Brajasundar Garnayak

Odisha, situated on the Bay of Bengal, is a part of the Eastern Ghat geomorphic province of India. The state is rich in archaeological vestiges from prehistoric times to the recent past. The early historic archaeology of the state shows a unique distribution pattern. More than dozens of early historic fortified settlements are found around the Eastern Ghat hill range as well as in the bank of Mahanadi and its tributaries. The richness of the region is credited to its strategic location between northern and southern India. The major ancient trade routes have passed through this land. The movement of people from the Ganga valley, Chhotanagpur plateau and from the Vindhyan region during the 1st and 2nd millennium BC, and the advancement of new knowledge of iron metallurgy had an immense impact on the socio-economic milieu in the state. The impact can be accessed from the excavated findings of early historic sites conducted in the last two decades.

The knowledge of iron smelting and depository of iron ore in the northern and central regions was known prior to the 1st millennium BC to the indigenous people of the state. The state is rich in the distribution of iron ore, having five distinct geographical zones, such as Bonai-Keonjhar, Gandhamardan, Tomaka-Daitari, Gorumohishani-Badampahar and Hirapur.[1] It possesses high-grade iron ore with an average content of iron over 60%. The deposits of iron ore are mainly found in Keonjhar district, followed by Sundargarh, Cuttack and Mayurbhanj districts.[2]

Beginning of Iron Technology

The development of iron smelting technology in the Indian Subcontinent is a debatable issue to date. Until the 1960s, many archaeologists believed that Indian iron smelting technology did not develop indigenously but is a result of diffusion from the west. In support of this diffusionistic theory, Allchin and Allchin in 1968 stated that knowledge of iron smelting technology was for the first time introduced in Asia Minor between 1860 BC and 1200 BC and subsequently spread towards the east and west.[3] L. Gopal in 1960 has mentioned that the Indian iron smelting technology was introduced by the Aryans people.[4] Supporting this view, N. R. Banerjee has mentioned that it

DOI: 10.4324/9781003370642-5

was the Painted Grey Ware (PGW) people who brought this technology to India. He has also argued that during their journey, the Aryans had picked this iron smelting technology from Iran.[5]

These diffusionistic views regarding iron technology were challenged subsequently. D. K. Chakrabarti argued for an indigenous development of iron smelting technology in India rather than diffusion from West Asia. However, Chakrabarti showed, on the basis of radiocarbon dates of some central Indian sites, that iron was produced in central India earlier than in the northwest.[6] The recent investigations carried out in the Gangetic plains in northern India have pushed back the introduction of iron technology to about 1800–1600 BC.[7,8]

The introduction of iron in the material culture of Odisha during the late Chalcolithic period had acted like a bridge between prehistory and history. The occurrence of iron in a presumably late Harappan context at Lothal and in an apparently mature Harappan context at Allahdino suggests that knowledge of iron smelting could have been a part of the Harappan tradition.[9] Iron technology evolved from the Chalcolithic metallurgical knowledge without any hiatus in Odisha. Most of the cultural components are the same, and only the presence and absence of iron is considered for determining the cultural phase as the Chalcolithic and Iron Age. The excavator of Golbai-Sasan, after noticing the similarity in material culture, termed the iron-bearing cultural horizon as Ferro-Chalcolithic (Period IIB dated to c.1100–900 BC of Golbai-Sasan). For the first time, the excavation at Golbai-Sasan has shown the evidence of a succession of Iron Age culture over the Chalcolithic culture.[10] The earliest evidence for iron comes from the excavated site of Kumersingha (Period-IA) 1,418 ± 223 BC and 1,648 ± 46 BC.[11] The excavations at various sites in the last two decades throw light on the beginning of iron metallurgy prior to the 1st millennium BC and had a distinct cultural horizon in the state.

Evidence of Iron Smelting

The evidence of iron artefacts, ores and slags has been reported from different sites in Odisha. The earliest evidence for iron smelting comes from the excavated site of Kurumpadar, located on the right bank of the Mahanadi. Significantly, from Kurumpadar, at a depth of 1.05 m, the evidence for iron smelting activities was noticed by the excavator. The floor which was partly burnt was found associated with charcoal fragments, ashes, lumps of iron slags and a complete specimen and two broken parts of tuyere.[12] Further evidence has been reported from Sankerjang;[13,14] Dangala and Gothapada in Gopalpur area of Nayagarh district Gopalpur;[15,16] Khambeswarpali;[17] Nuagarh-Asurgarh;[18] Bharati Huda;[19] Kapasira, Khajeriapali, Amudda, Kurumpadar, Subulia and Badmal in the Mahanadi valley;[20] Amdeah in Sambalpur district;[21] Jiragoara and Chiglabecha areas of Koraput;[22] Nadangipenth in Jajpur;[23] Guapada and Betgarh on Karandi river valley in Angul and Sambalpur districts;[24] Kharligarh and Gudavella in Bolangir district;[25,26] Sisupalgarh;[27] Golbai-Sasan in Khurdha district;[28] Jaugarh;[29] Asrubahal, Belparha, Chandupal Dahamahal, Dubanali, Durgapur Kandha Sahi, Durgapur, etc. in central Odisha;[30] Balitangri, Baghia, Bijatala, Bagdega, Tirilipi, Auski, Bhatchatra, Rengalbera, Telkoi, etc. in Mayurbhanj and Keonjhor districts of northern Odisha[31] and Chandikhal.[32] The above-mentioned sites provide a clear picture about the distribution and knowledge of iron technology. Iron metallurgy started in the state prior to the 1st millennium BC and continued till the pre-industrial era, having a climax during the 3rd century BC as observed from the iron tools retrieved from the excavations. The availability of rich iron ore deposits in the central and northern parts of Odisha possibly paved the beginning of early metallurgy in this land.

Antiquity of Iron from Radiocarbon Dates

The radiocarbon dates from different excavated sites collected from the iron objects bearing strata suggest that Odisha was one of the regions in ancient India having rich and earliest evidence of iron metallurgical traditions. The earliest evidence for iron comes from the excavated site of Kumersingha (Period-IA) at 1,418 ± 223 BC and 1,648 ± 46 BC.[33] Besides, a series of dates prior to the 1st millennium BC to the 6th–7th century BC have been reported. The excavation at Asurgarh from 2018 to 2019 revealed iron objects from the beginning of the settlement, i.e. Period I dated to 888 BC and 778 BC (Mean C^{14} date in BC). The finest variety of iron objects yielding from Period II in the site dated to 340 BC (Mean C^{14} date in BC).[34] In the Middle Mahanadi valley, the excavated sites of Khameswaripali-Period-IB dated to 1,065 ± 66 BC; Badmal-Asurgarh (Period-IB) 1,319 ± 132 BC and 838 ± 86 BC; Kumersingha (Period-IA) 1,648 ± 46 BC; Kapasira (Period-IA) 875 ± 475 BC; Kantipuleswar (Period-IA) 855 ± 125 BC and Kantipuleswar (Period-IA) 1,180 ± 210 BC have yielded iron objects associated with the similar type of ceramics as noticed in the late Chalcolithic phase.[35] For example, Golbai-Sasan (Period IIB dated to c.1100–900 BC), Bharati Huda (Period II) 1072 BC, shows the continuation of earlier cultural features except the introduction of pre-firing painted pottery. As noticed from these sites, the evidence of Iron Age cultural level gradually developed from the Neo-Chalcolithic cultural horizon, and as in Hikudi, the Iron Age succeeded the Neolithic period;[36] in Khameswaripali, Bharati Huda, Golbai-Sasan, it originated from Chalcolithic level. In general, the presence and absence of iron is the only differentiating factor for designating the cultural phases in Odisha as Chalcolithic or Iron Age.

Iron Technology and Urbanization

The beginning of iron technology and its impact on the socio-economic milieu in Odisha is well reflected in the excavated findings. The spread of iron technology in the state prior to the 1st millennium BC played a major role in the beginning of early urbanization. There are hundreds of sites that flourished during that time in the major and minor rivers' valleys. Most of the sites are well fortified and strategically located and have fulfilled the concept of *Durga* according to the *Arthasastra* of Kautilya. The ancient urban centres were located either near the fertile track or near the resource centres. Sites such as Badmal Asuragrh, Narla-Asurgarh and other Asura sites in the Mahanadi valley and its tributaries are located near the semiprecious gemstone quarries.

The sites like Jaugad, Lathi, Radhanagra, Sisupalgarh in the eastern region are located in fertile land tracts. The introduction of iron technology in these areas drastically influenced the sectors of agriculture and trade-commerce and settlement patterns as well. The production of agricultural resources increased due to inputs of iron technology which paved the way for trade and commerce. As a result, it led towards the beginning of urbanization (Figure 4.1). The high-quality iron tools of war and peace retrieved

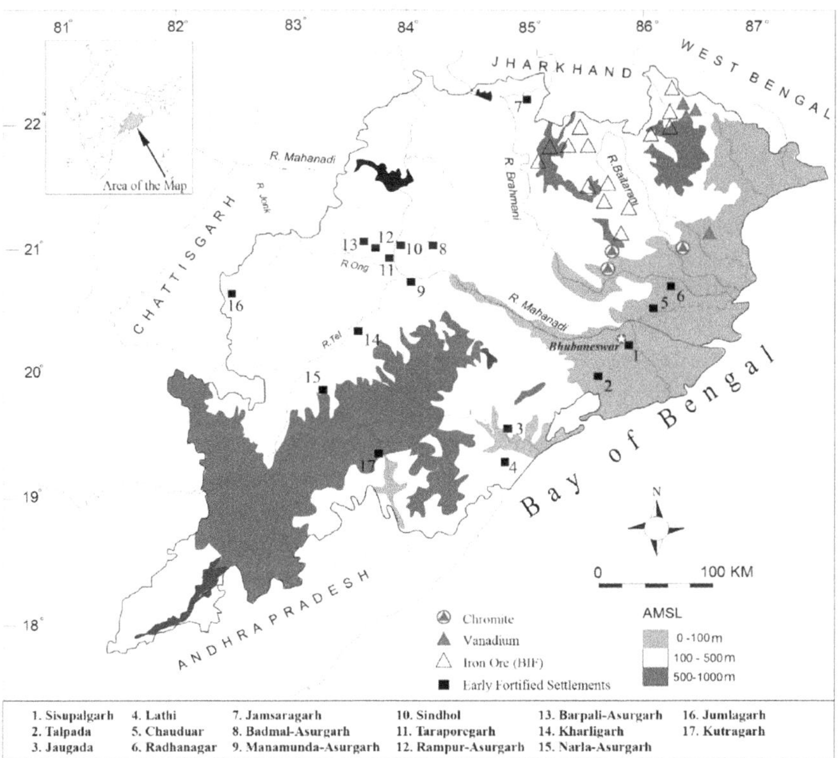

Figure 4.1 Map Showing the Locations of Ancient Urban Centres and Iron Ore Quarry Sites. Source: *Byomakesh Tripathy and Dibishada Brajasundar Garnayak*

from Kharligarh and Narla-Asurgarh have shown advanced iron metallurgy in the 3rd–4th century BC. Some of the iron tools have not been rusted to date, and even a few objects are ochre coated. The findings of different shapes and sizes of iron objects from the excavations of other sites also show the richness in variation. The most common iron objects found at the sites are harpoons, spearheads, arrowheads, knives, chisels, nails, clamps, sickles, etc. It appears from the above evidence that specialized craft industries were introduced during ancient times, which laid the foundation for the first phase of urbanization in Odisha during the 3rd–4th century BC in particular, whereas the second phase of urbanization in the subcontinent evolved a little late.

Gradual Evolution of Urbanization in the Context of Iron Technology

The urban centres found in the form of well-fortified settlements in the Mahanadi valley have rich legends and traditions. The locals believe that Asuras of epic times were the builders of such huge fortified settlements. In the case of Narla-Asurgarh, the locals assign the fort to Gosingha-Dutta, the son of the demon Mahisasura of epic time. The beginning of the Iron Age and early historic sites in Odisha has a strong affiliation with Asura sites. Besides, a host of Asura sites are found in the neighbouring states of Chhattisgarh, Jharkhand and West Bengal. The material cultures retrieved from these sites are of burnt bricks, iron implements and slags, potsherds, stone implements, cinerary urns, silted-up tanks, etc. In Odisha, the Asura sites are Narla-Asurgarh, Manamunda-Asurgarh, Kharligarh, Badmal-Asurgarh, Barpali-Asurgarh, Nuagarh-Asurgarh, etc. These sites are locally known as Asurgarh on account of the occasional discovery of human skeletal remains, iron objects and potteries of unknown antiquity in the cultivated surface and eroded sections of the mound. The use of iron, punch-marked coins and potteries of black slipped and black-and-red ware remains the hallmark of the material assemblage of these sites. Most of the Asurgarh sites were fortified metropolises, where large-sized bricks had been extensively used for construction. The sites were strategically located in the Mahanadi valley and along with its major tributaries. The excavation in most of these sites revealed that these sites were abandoned and deserted after the 3rd–4th centuries AD. It is also interesting to note that an iron smelting tribe called the Asuras is found in the Ranchi district,[37] but their connection with the Asuras of the Munda tradition is not clear.

Whatever be the tradition of iron smelting and metallurgy, it was a full-fledged occupation of a few ethnic groups in Odisha and its neighbouring states. During the pre-industrial phase, the Kamar/Luhar/Agaria/Asur community prepared tools (sickle, hoe, ploughshare, axe, pick-axe, etc.) for using them in agricultural fields. The tradition of iron smelting of Odisha and its neighbouring state is still seen in rural areas. The

Agaria community, residing in the neighbouring state of Chhattisgarh, is indigenous iron smelters and craftsmen of the tribal community who use the indigenously developed technique of iron smelting and craft work to produce iron from locally available iron ore to make artefacts and agricultural implements.[38] Regarding the iron smelting activities, some ethnographic data have been collected from village Bhagabandhi, close to the rich iron ore deposit of Badampahar in northern Odisha by B. K. Mohanta, that throws light on this traditional technique of iron smelting in the colonial era.[39]

Metal always plays a significant role in the material culture of any society. Iron is one of the materials which brought changes in the course of history and shaped the formation of many states and dynasties in early historic times. The basic origin and growth of urban centres in Odisha was due to the knowledge of advanced iron metallurgy by the inhabitants followed by landscape potentiality and richness in mineral resources. The gradual development of iron technology in the state and the movement of people and culture due to trade and commerce in the early historic times must have paved the way for first urbanization. The Iron Age settlements have their beginning prior to the 1st millennium BC and continued till the 2nd–3rd century AD. Also, it has been observed that the period 200 BC–AD 100 marked progress and richness in the material culture. This is the period which shows the first urbanization in this land of Odisha when North India witnessed the development of second urbanization. Such a type of revolution was never seen in the history of Odisha before and after.

Notes

1 D. K. Chakrabarti, *The Oxford Companion to Indian Archaeology*, Oxford University Press, New Delhi, 2006, pp. 287–302.
2 P. K. Behera, et al., "Iron Age in the Middle Mahanadi Valley of Odisha, India, Recent Perspectives", *Heritage, Journal of Multidisciplinary Studies in Archaeology*, Vol. 7, 2019, pp. 95–134.
3 Bridget Allchin and Raymond Allchin, *The Birth of Indian Civilization*, Penguin Books Ltd., Harmendsworth, 1968, pp. 207–232.
4 Lallanji Gopal, "Antiquity of Iron in India", *Uttara Bharati*, Vol. IX, (3), 1960, pp. 71–80.
5 N. R. Banarjee, *The Iron Age in India*, Munshiram Manoharlal, Delhi, 1965.
6 D. K. Chakrabarti, *The Beginning of Iron in India*, *Antiquity*, Cambridge University Press, 1976, L, pp. 114–124.
7 R. Tewari and R. K. Srivastava, *Excavations at Raja Nala-ka-tila* (1996–1997), District Sonbhadra (Uttar Pradesh), Preliminary Observations, *Pragdhara*, Lucknow, 1998, Vol. 8, pp. 99–105.
8 R. Tewari, et al., *Excavation at Lahuradewa*, District Sant Kabir Nagar, Uttar Pradesh, *Puratattva*, New Delhi, 2002, Vol. 32, pp. 54–62.
9 D. K. Chakrabarti, *The Oxford Companion to Indian Archaeology*, *op. cit.*, pp. 287–302.
10 B. K. Sinha, *Excavations of Golabai-Sasan*, District Puri, Orissa, *Puratattva*, New Delhi, 1992–1993, No. 23, pp. 48–50.
11 P. K. Behera, et al., *op. cit.*, pp. 95–134.

12 P. K. Behera, *Excavations at Kumersingha and Kurumpadar – The Iron Age Settlements in the Middle Mahanadi Valley, Orissa: Results of the First Season Work. Pragdhara*, 2002–2003, Vol. 13, pp. 87–103.
13 Paul Yule and B. K. Rath, "*Sankerjang - A Metal Period Burial Site in the Dhenkanal Uplands of Orissa*", in K. K. Basa and P. Mohanty (eds.), *Archaeology of Orissa, Pratibha Prakashan*, New Delhi, 2000, Vol. 1, pp. 285–321.
14 R. N. Dash, "Sankerjang - A Chalcolithic Site in Orissa", *Orissa Historical Research Journal*, Bhubaneswar, Vol. 32, 1986, pp. 99–125.
15 S. K. Kar, *Further Exploration at Gopalpur, Puratattva*, New Delhi, 1995–1996, 26, pp. 105–106.
16 S. K. Kar, "Gopalpur: A Neolithic-Chalcolithic Site in Coastal Orissa", in K. K. Basa and P. Mohanty (eds.), *op. cit.*, pp. 368–391.
17 P. K. Behera, *Excavations at Khameswaripali - A Protohistoric Settlement in the Middle Mahanadi Valley, Odisha: A Preliminary Report, Pragdhara*, 2000–2001, Vol. 11, pp. 13–34.
18 S. Pradhan, "Excavation at Nuagarh-Asurgarh: A Preliminary Report", in S. Pradhan (ed.), *Art and Archaeology of Orissa: Recent Perspectives*, Aryan International, New Delhi, 2006, pp. 63–79.
19 D. B. Garnayak, "Excavation at Bharati Huda a Protohistoric Settlement in the Mahanadi Delta, Eastern India", in *Puranveshana*, New Delhi, 2019, Vol. 3, pp. 1–14.
20 P. K. Behera, et al., *op. cit.*, pp. 95–134.
21 D. K. Chakrabarti, *op. cit.*, pp. 287–302.
22 G. Kuppuram, *Ancient Indian Mining, Metallurgy and Metal Industries*, Sundeep Prakashan, Delhi, 1989.
23 S. Acharya and S. Acharya, "An Indigenous Blast Furnace of Iron Smelting at Nadangipenth, Daitari Hills", in *Proceedings of the 48th Session of the Indian History Congress*, held at Goa, 1987.
24 A. Pradhan, *Exploration in Districts of Angul and Sambalpur*, *Indian Archaeology: A Review*, New Delhi, 1996–1997, pp. 70–76.
25 R. N. Dash, "The Iron and Other Objects from Gudavella and Kharligarh", *Journal of the Orissa Research Society*, Vol. 1, No. 2, 1982, pp. 53–61.
26 D. B. Garnayak, *Early Historic Archaeology of Odisha*, Mayur Publications, Bhubaneswar, 2012, pp. 30–53.
27 B. B. Lal, *Sisupalgarh 1948: Early Historic Fort in Eastern India*, 1949, Ancient India-5, New Delhi, pp. 62–105.
28 B. K. Sinha, *op. cit.*, pp. 48–50.
29 D. Mitra, *Excavations at Jaugarh in the District of Ganjam, 1956–57, Indian Archaeology: A-Review*, New Delhi, pp. 30–31.
30 S. K. Kar, *Archaeology of the Singida Valley Central Orissa*, Unpublished PhD Thesis Submitted to Utkal University, 2008, Bhubaneswar.
31 B. K. Mohanta, "Pre-Industrial Iron Smelting Technique of Eastern India: with Special Reference to Odisha", in B. Tripathy and D. V. Prasad (eds.), *Ethnoscience and Traditional Technology in India*, Aayu Publication, New Delhi, 2016, pp. 55–80.
32 D. Sahoo, *An Ancient Iron-Smelting Site in Coastal Orissa, Man in Society*, Bhubaneswar, 2001–2003, 14 and 15, pp. 77–93.
33 P. K. Behera, et al., *op. cit.*, pp. 95–134.
34 D. B. Garnayak, *Excavation at Narla-Asurgarh*, Unpublished Report of Archaeological Survey of India, New Delhi, 2019, pp. 12–14.
35 P. K. Behera, et al., "Radiocarbon Dates from Middle Mahanadi Valley, Odisha", *Man and Environment*, Vol. XXXII, No. 2, 2007, pp. 41–46.
36 P. K. Behera, "Investigations into the Mesolithic and Post-Mesolithic Settlements in the Middle Mahanadi Valley, Orissa", in S. Pradhan (ed.), *op. cit.*, pp. 1–62.

37 K. K. Leuva, *The Asur: A Study of the Primitive Iron Smelters*, Bharatiya-Adimjati-Sevak Samaj, New Delhi, 1963, pp. 44–62.
38 Amit Soni, "Iron-Metal Craft Economy of the Agaria Tribe of Chhattisgin", in B. Tripathy and D. V. Prasad (eds.), *op. cit.*, pp. 370–376.
39 B. K. Mohanta, *op. cit.*, pp. 55–80.

5 The Aryabhata Legacy

Sabyasachi Chatterjee

Introduction

On the question of India's contributions to the world of science, it is recognized that "the foundations of modern arithmetic and algebra were laid long ago in India". Further,

> these advances in mathematics are contained in books written by a succession of eminent mathematicians from the 5th to the 12th century AD. But the earliest extant book on algebra is by the famous astronomer, Aryabhata, who was born in AD 476.[1]

The importance of India's contribution lay in giving

> foundational departures in mathematics (particularly in the development and use of the decimal system and in trigonometry) and also to new ideas and measurements in astronomy. Indeed, the departures presented in Aryabhata's pioneering book, completed in 499 AD, not only generated extensive responses within India (starting with the works of Varahmihira, Brahmagupta and Bhaskara), but they were also much discussed in their Arabic translations.[2]

Notwithstanding the contributions of earlier researchers, the Indian tradition of astronomy and mathematics was pioneered by Aryabhata (AD 476–AD 550). As Rao observes, the arrival of Aryabhata resulted in the replacement of the earlier "astronomical computations" described in the *Vedangajyotisa* with a "new type of Indian astronomical literature" called *Siddhanta* having connotation of an established theory.[3] Importantly, "the astronomical parameters and formulas given by Aryabhata were not merely theoretical manipulations or speculative conjectures. These results were based on actual observations and verifications".[4]

By naming India's first spacecraft after Aryabhata, launched on 19 April 1975, the nation had acknowledged the pioneering role that Aryabhata had played in laying the foundation for the mathematical and astronomical

DOI: 10.4324/9781003370642-6

tradition of India. It should, however, be recorded that the scientific inputs that went into the design, construction or launching of the spacecraft had no relation with the mathematical formulations that Aryabhata had made. This last point has to be reiterated since interested parties in recent years publicize the idea, often from platforms of governance and administration, that modern science can be rediscovered from "Hindu" spiritual books of antiquity. Such wasteful exercise of trying to learn modern science from ancient books has to be discouraged, and fallacious claims have to be exposed.

As regards the Indian mathematical tradition, some confusion can arise with the term "Vedic Mathematics". This term got currency with the publication of the book, of the same title, in 1965, authored by a religious figure, Jagadguru Swami Shri Bharati Krishna Maharaj. The title suggests that the mathematics expounded in the book originated in the antiquity of Vedic times. On scrutiny, researchers have found that "an analysis of the mathematical contents of the book also shows they cannot be from the Vedas" and that "when the superimposed "Vedic" garb is taken off, the material in the Swamiji's book looks like a compilation of tricks in elementary arithmetic and algebra, discovered by some interested hobbyists".[5]

Similarly, if the ancient Indians had such a developed knowledge of plastic surgery, why was its usage not cited in the *Ramayana* or *Mahabharata*? In the wars that are described in these classics, hundreds lay maimed. Some were humans, some were horses and others were elephants. We get no reference as to how much such inter-species transplantation took place at that time! What is even more surprising is that when India's former Prime Minister's knee surgery had to be performed, it was not done according to the prescriptions given in the ancient Hindu texts, but a surgeon of Indian origin, practising in the USA, had to be transported to India. All his surgery was based on the knowledge from the West, and he is not known to have done the surgery by consulting any of the ancient Hindu texts.

The point that has to be emphasized is that the claims about achievements have to be done on the basis of objective reality. The tragedy is that nowadays false claims are projected as true, and actual achievements are being side-lined.

In making an objective assessment about Aryabhata, one must keep in mind that the commentaries that we have on Aryabhata's research findings are based not on the original manuscripts but are derived from the writings that have been passed down from records of the original. These are based on the manuscripts that are available in different places, e.g. in the Kerala University of Oriental Research, Thiruvananthapuram, at the Oriental Government Manuscripts Library, Chennai, and in private collections of several families. The critical edition has been prepared based on several palm-leaf texts in Malayalam script that are available at the Kerala University of Oriental Research, Thiruvananthapuram. "All these manuscripts were completely independent of each other" and with minor variations.[6]

The author of the critical edition also compared the text preserved in the commentaries of *Aryabhatiya* as well as extracts from it that occur as quotations of later astronomers' works. It was noted by them that some of these Suryadeva's commentaries on Aryabhata were "replete with all types of scribal and intrinsic errors, including omissions, gaps, wrong prescriptions, textual corruptions, etc. and so was utterly unreliable",[7] and were, hence, discarded in their researches on Aryabhata. In the selection of manuscripts, scholars relied on the textual similarities of the manuscripts, such that it was surmised by them that continuity amongst these manuscripts implied that they were derived from the same source and carried minimal discrepancy between them.

As is known, Aryabhatan School flourished in Kusumapura, the present-day Patna, as is declared in *Aryabhatiya* itself.[8] Said to be born as an Asmaka, Aryabhata was thus an outsider to Magadha.[9] As is the case with many scholars with immense enthusiasm for learning, Aryabhata thus seems to have left his Asmaka roots to absorb the knowledge from the land of Kusumapura. In Kusumapura, he was regarded as a "*kulap*", i.e. as a teacher in an educational institution.[10] While some have speculated that this institution belonged to Nalanda University,[11] others are doubtful of this Nalanda lineage.[12] They say, had Aryabhata been a teacher in Nalanda, he would have directly declared the Nalanda University to be the patron rather than acknowledging Kusumapura. In my opinion, this particular aspect of Aryabhata being a *kulap* needs careful scrutiny. This would imply that there existed a school, and therefore, it is possible that the work in *Aryabhatiya* was a compendium of works done by several researchers involved with that school. Given the great prestige that Nalanda had in the world of scholarship, it would also be important to investigate the interactions between Kusumapura and Nalanda; Nalanda being the centre to which many scholars would congregate from different parts of the world.[13]

The mathematical and astronomical works of *Aryabhatiya* appear to be of such range and richness that Bhaskara I (7th century AD) considered that Aryabhata's achievements could only be attributed to divine gifts beyond the reach of senses – "His grace was revealed to him the true knowledge of the subject pertaining to the true motion of planets".[14] Such myths also exist about Kalidasa (belief, that his creativity was born due to blessings from Saraswati, the Hindu goddess of learning!), and about the 20th-century mathematical genius Srinivasa Ramanujan, who was believed to be guided and led by goddess Namagiri.[15] Though Bhaskara I had also considered the possibility that Aryabhata derived these results on the basis of observation and inference, the idea of divine interference took precedence over that of these worldly exercises of observation and inference.[16] An alternative version, according to the present author, is that the *Arybhatiya* may be the contribution from Kusumapura School which Aryabhata founded, in the same way that Pythagorean School's contributions are often considered, by the layman, to be the achievement of a single individual. This needs to be investigated as Bhaskara I does refer to the "learned people of Kusumapura"[17] and further emphasized that academies of advanced learning did exist there, and Aryabhatan School was possibly one such.

Mathematical Contributions of Aryabhata

Aryabhata's contributions lie in the areas of astronomy, questions of time reckoning and spherical geometry and trigonometry. We shall focus our attention on two of the most important results that appear in *Aryabhatiya*. These pertain to the following:

i. The value of π (the ratio between the circumference and diameter of a circle)
ii. The Rsine tables

In what follows, we will briefly describe them and give our comments on the values described in *Aryabhatiya*.

On the Value of π (The Ratio between the Circumference and Diameter of a Circle)

Aryabhatiya gives the following prescription. That the ratio between the circumference and the diameter of a circle is constant was known for long. In the Bible, it is stated that this number is 3, presumably taken from empirical methods. In schools, we are taught that this number is (22/7), a value obtained by Archimedes by dividing a circle into a regular polygon with 96 equal sides. Archimedes had concluded that the value of this ratio lies between (22/7) and (223/71) and that the first value is taken for the convenience of calculations. This ratio was referred to as π by Euler in 1737, and this symbol is now universally used to refer to this ratio.[18] Aryabhata had given its value by the following *sutra:*

> 100 plus 4 multiplied by 8 and added to 62000: this is the nearly approximate measure of the circumference of a circle, whose diameter is 20000.[19]

According to the above *sutra*, the π = circumference/diameter = $[\{(100+4) \times 8\} + 62000]/20000 = 62832/20000 = 3.1416$.

If one notes the value of π as given in Clarke's tables, which every school student is familiar with these days, the value is found to be 3.14159, i.e. Aryabhata's value had an error of +0.00001, which is an error of $[(+0.000001/3.14159) \times 100] = +0.000318\%$, i.e. only 3.18 in one million.

This was a great achievement, though it was not the most accurate for the time. The 5th century AD Chinese mathematician and irrigation engineer, Zhu Chongzi, got a result that the value of π lies between (22/7) and (355/153), i.e. more accurate than Aryabhata's when compared to the value known in modern days.[20] In the 16th century AD, the Persian astronomer, Jamshed al Kashi, working in Samarkand, found the value correct up to the 16th place of decimal which no mathematician had succeeded till then.[21]

An intriguing question, however, arises. It is true that for the sake of accuracy, Aryabhata had put the value to be the ratio between two large integers. But on removing the common factor of 16, between the numerator and denominator, it could as well have been expressed as (3927/1250). So why such two very large integers like 62832 and 20000 were chosen is not clear.

It is to be noted that the number π = (62832/20000) was found to be acceptable to the mathematician community, possibly for the convenience of calculation. Thus, we find that the famous Persian mathematician Mohammed al Khwarizmi (c. AD 780–c. AD 850) states,

> In any circle, the product of its diameter, multiplied by three and one-seventh, will be equal to its periphery. This is the rule generally followed in practical life, though it is not quite exact. The geometricians have two other methods. One of them is that you multiply the diameter by itself, then by ten, and hereafter, take the root of the product; the root will be the periphery. The other method used by astronomers among them is that you multiply the diameter by sixty-two thousand eight hundred and thirty-two and then divide the product by twenty thousand; the quotient is the periphery. Both methods come very nearly to the same effect.[22]

Thus, according to the first prescription, we would have $\pi = \sqrt{10} = 3.1622$, while the second one is a far more accurate value, i.e. $\pi = 3.1416$, as was given by Aryabhata. Thus, the knowledge about this value was transmitted to Persian and Central Asian Schools well before Al-Beruni's *Kitab-al-Hind* was published.

In Aryabhata's statement, about the value of π, he had stated that π = (62832/20000) was an approximate one. He qualified his value of 3.1416 for π as "âsana", "approximate or approaching". On using the Archimedes method, i.e. dividing a circle into a regular polygon of 512 sides, it has been found that this method gives the value of π to lie between 3.14157 and 3.14163. This was the method that was used by the present author using a programme written with the Octave. On averaging, the value is 3.1416, which Aryabhata gave. We can thus state that it is quite possible that Aryabhata had used the Archimedes method to find the value of π that is attributed to him. As *Aryabhatiya* gave no indication about the method, the above may be a possible one, though it is not possible to state the level of computational accuracy that was aimed at and achieved in order to arrive at the above value.

Rsine Tables

The Rsine tables, given by Aryabhata, give the difference, Rsine = R [sin (θ (n+1)) – sin (θ (n))], where [θ (n)] = n × 3.75 degrees [3.75 degrees = 3°45′ of arc]. From Aryabhata's Rsine tables, it follows that he had computed the Rsine's by taking R = 3438. In the following Table 5.1, we list the values as given in *Aryabhatiya* and compare them with the values that we obtain with a simple digital calculator, both being taken with R = 3438.[23]

It is seen that the Rsine values, i.e. the sine differences prescribed by Aryabhata, were extremely accurate when tested with the values known today by digital computers. Accuracy in the actual sine values is even higher.

An inconsistency, however, remains. The Rsine values given above do not match with the value of π = 3.1416 but rather to π = 3.16. This inconsistency has to be investigated. It is to be recognized that π is a number that finds application in various fields, for example,

> it is the basis of the cyclometer (the first model of which was made in Alexandria about 100 BC) and the speed indicator. It is the basis of all large earth measurements and estimates of the sun and moon. As we know the circumference of the Earth, it gives us the Earth's radius and the circle of any circle of latitude. Without π there would be no Columbus and no George Stevenson.[24]

Aryabhata's Legacy: Acknowledgements and Criticisms

Aryabhata was a pioneer of his time. Future astronomers and mathematicians like Varahmihira (6th century AD), Bhaskara I (7th century AD) and Brahmagupta (c. 7th century AD) have acknowledged it. However, there are also criticisms of the accuracy of many of the predictions of Aryabhata, notably by Brahmagupta and Lalla (c. 8th century AD). The latter had commented that the prescriptions, as given by Aryabhata and his followers, were often not clearly stated, and so Lalla had taken up the task of restoring them. A question will arise: why were Aryabhata's mathematical results written in incomprehensible manners, as alleged by Lalla? Was that done in order that the knowledge did not leave outside the school the same way that the Pythagoreans used to practise? With the southward advance of the Magadha Empire, mathematical and astronomical knowledge too travelled southward.

Copious references to Aryabhata appear in several sources that appear in far-flung places and in different eras. These are Varahmihira (Ujjain, late 6th century AD), Bhaskara I (Kathiawar, 7th century AD), Prthudaka (Kannauj, UP, 7th century AD), Brahmagupta (Rajasthan, c. 7th century AD), Bhattapala (Kashmir, late 10th century AD), Suryadeva (Thanjavur, Tamil Nadu, 13th century AD), Lalla (c. 8th century AD), Yallaya (Srisaila, Kurnool district in Andhra Pradesh, late 16th century AD) and Neelakantha

Table 5.1 Rsine Tables

n	$\theta(n)$	*Rsine (Aryabhata's)*	*Rsine (Modern)*	*Relative % Error in Rsine*	*Rsin $\theta(n)$ (Aryabhata's)*	*Rsine $\theta(n)$ (Modern)*	*Relative % Error in Rsine $\theta(n)$*
0	0	0	0	0	0	0	0
1	3°45′	225	224.8560	+0.0640	225	224.8560	+0.0640
2	7°30′	224	223.8930	+0.0477	449	448.7490	+0.0590
3	11°15′	222	221.7315	+0.0128	671	670.7205	+0.0416
4	15°00′	219	219.0994	−0.0454	890	889.8199	+0.0202
5	18°45′	215	215.2890	−0.1340	1105	1105.1089	−0.00985
6	22°30′	210	210.5567	−0.2640	1315	1315.6656	−0.0506
7	26°15′	205	204.9229	+0.0376	1520	1520.5858	−0.0387
8	30°00′	199	198.4115	+0.02960	1719	1719.0000	0
9	33°45′	191	191.0504	−0.0264	1910	1910.0504	−0.0026
10	37°30′	183	182.8714	+0.0703	2093	2092.9218	−0.0037
11	41°15′	174	173.9091	+0.0523	2267	2266.8309	−0.0074
12	45°00′	164	164.2022	−0.1230	2431	2431.0031	−0.0013
13	48°45′	154	153.7921	−0.1350	2585	2584.8252	−0.0068
15	52°30′	143	142.7236	-0.1930	2728	2727.5488	+0.0165
16	56°15′	131	131.0437	-0.0300	2859	2858.5925	+0.0142
17	60°00′	119	118.8026	+0.166	2978	2977.3953	+0.0203
18	63°45′	106	106.0532	−0.0501	3084	3083.4485	−0.0179
19	67°30′	93	92.8493	−0.1630	3177	3176.2978	+0.0221
20	71°15′	79	79.2480	−0.3130	3256	3255.5488	+0.1390
21	75°00′	65	65.3042	−0.4660	3321	3320.8530	+0.0044
22	78°45′	51	51.0868	−0.1700	3372	3371.9398	+0.0018
23	82°30′	37	36.6476	+0.9610	3409	3408.5874	+0.0121
25	86°15′	22	22.0518	−0.2350	3431	3430.6390	+0.0105
26	90°00′	7	7.3610	−3.0000	3438	3438.00	0

Somiyaji (Kerala, early 16th century AD). Many important results are contained in *Aryabhatiya* and *Aryabhatta-siddhanta*. The last work has not survived, but the results can be traced in the commentary of the same that Brahmagupta wrote in the c. 7th century AD "*Khandakhadyaka*". The last work was translated into Arabic by Al-Beruni, in the 11th century AD.[25]

It has to be noted that Brahmagupta's references to Aryabhata were not always adulatory, and there are strong strictures too. In the *Uttarakhandakhadya* part of *Khandakhadyaka*, Brahmagupta writes, "I now give corrected rules as Aryabhata's formulas do not give quite accurate results"; in other words, they do not tally with the observations.[26] Similarly, Lalla had expressed the following criticism of Aryabhatan School,

> Lalla now depicts the motion, etc. of the heavenly bodies, a subject which has been dealt with by Acarya Aryabhata. But the subject is rather difficult ... It is true that Aryabhata's pupils have made complete study of the works of their teacher and have written treatises on astronomy; but they have not dealt fully with the procedures of computations. Therefore, I shall deal, in proper order, with the procedures of what has been enunciated by Aryabhata.[27]

This means that though Aryabhata was acknowledged as a pioneer, his "prescriptions" were not accepted uncritically. Thus, elements of criticism and debate were ingrained in the ancient Indian academic tradition. Indians were argumentative and have remained so.[28]

A distinct hallmark of *Aryabhatiya* is that in his work, Aryabhata followed a critical approach. This is amply demonstrated in Aryabhata's *sutra* regarding the measurement of π, where he had commented that the value of $\pi = (62832/20000)$ is an "approximate" one. In other words, it acknowledged the limitation of his value of π and invited others to find the better ones. As we know, Aryabhata had stated the results, but the method was not elaborated. It is not possible to determine the reason, but such intended secrecy could have become an impediment to the further development of Kusumapura School. Lack of transparency could have led to speculations and errors that Lalla sought to correct. It is heartening to note that this type of obscuration was not followed by the later scholars of the South. For example, Suryadeva describes, in great detail, the construction of a *golayantra*, i.e. a chronometer which, by regulating the flow of water, keeps a sphere rotating at a speed of half a rotation in 12 hours.[29]

Myths about Aryabhata: Bihar, Asmaka or Kerala?

Concerning Aryabhata and his school, certain questions have to be investigated. For how long did this mathematical school flourish in Kusumapura? Why did it disappear from there while it spread to other parts of India, presumably with the expansion of the Magadha Empire? Another question concerns his place of birth. Bhaskara I says that Aryabhata was an Asmaka, so he was an outsider to Kusumapura but settled down and established himself in Kusumapura. The social factors that extended this patronage to this astronomical and mathematical school should also form an important subject of study.

While there is a consensus that Kusumapura, where Aryabhata worked, lies in the present-day city of Patna, in recent times it has been claimed that Aryabhata belonged to Kerala.[30] In the interest of objectivity and scholarship, this controversy has to be investigated since this has an important bearing on the question of generation of the knowledge system in ancient India and also its preservation and dissemination. The idea that Kusumapura was located in Pataliputra would imply that the Kusumapura tradition travelled to the South with the southward expansion of the Magadha Empire. This southward transit germinated on the way to Andhra School and Thanjavur School in the 13th century, while Kerala received the knowledge later in the most mature form.[31]

Let us now make the reverse journey. We assume that Aryabhata worked in Kerala in the late 5th and early 6th century AD. Then this knowledge would have travelled northward where it was received by Al Biruni in the 11th century AD, and he would later take it to Afghanistan, Persia, Iraq and the rest of Arabia. By what process was this knowledge carried from Kerala to the North? How long did it take to travel, and who were the carriers? Did it travel from Kerala to Ujjain in less than a century to be picked up by Varahmihira? Further, what was happening in astronomy in Kerala between the 6th century and the time of Neelakantha Somayagi and his followers, i.e. the 15th–16th century AD? That is, what was the state of astronomy in Kerala in the intervening period; was nothing happening there? In other words, if Aryabhata belonged to Kerala in the 5th century AD, then this knowledge jumped to Ujjain within one century, without leaving a mark anywhere in between, and then made a gradual southward progress to the South for about a thousand years to reach its pinnacle with Neelakantha Somayagi, i.e. back to Kerala again.

Conclusion

No matter what the final findings may be, Aryabhata's contributions to knowledge need to be evaluated from a multidimensional perspective. Laying the foundations of Indian astronomical knowledge was a singular contribution of Aryabhatan School. The route, by which it was transmitted in India and outside, needs a thorough and systematic study. There is a school of thought that India lives in the spiritual world, while the materialistic West is the repository of scientific knowledge. There is no doubt that spiritual ideas did have a wide circulation in Indian society and still do. But so it was for Europe also. In India, however, spiritual philosophies of different kinds have survived for centuries, but they have not lived without challenge. The important sociological issue is, in spite of such an early start, why did the Indian scientific institutions fail to achieve the breakthrough that western societies achieved? Was it because of the conservatism in the academia of the time, or was it the conservatism of the contemporary society that infiltrated into the institutions of learning?

The noted chemist and historian of chemistry, Dr. Prafulla Chandra Ray, has commented that the caste system contributed to the decline and collapse of chemistry in ancient India.[32] What was the cause of the decline of astronomy and mathematics in ancient India needs critical examination. It is known that social interactions between nations and cultures gave rise to the flourishing of science in many a society in ancient and medieval times. In these processes, contacts between India, China, Persia and the Arab world played very important roles.[33] The large question is this: while Europe became the recipient and beneficiary of science that was transmitted by the above societies, what led to the collapse of science in India? Simplistic answers that "foreign invasions" was the cause would not satisfy. The counterpoint would be why did the flowering of science in Europe follow foreign invasions, i.e. by Asiatic people? Objective evaluation of these questions will enhance the citizen's confidence in scientific temper, an issue that is directed by Article 51a–h of the Indian constitution.

Notes

1 Jawaharlal Nehru, *The Discovery of India*, Penguin Books, New Delhi, 2004, pp. 230–235.
2 Amartya Sen, *The Argumentative Indian*, Penguin Books, London, 2005, p. 28.
3 S. Balachandra Rao, *Indian Astronomy: An Introduction*, Universities Press, Hyderabad, 2000, pp. 6–10.
4 S. Balachandra Rao, *Indian Mathematics and Astronomy: Some Landmarks*, Bhavan's Gandhi Centre of Science and Human Values, 2004, p. 52.
5 S. G. Dani, "Vedic Mathematics - Myth and Reality", *Economic and Political Weekly*, Vol. 28, 1993, pp. 1577–1580.
6 Kripa Shankar Shukla (ed.), *Aryabhatiya of Aryabhata*, Part I, Indian National Science Academy (INSA), New Delhi, 1976, pp. lxviii–lxxii.
7 *Ibid.*, Part III, pp. xvii–xx.
8 *Ibid.*, Part I, pp. xvii–xix; p. 33.
9 *Ibid.*, Part I, pp. xvii–xix; p. xix; Part III, pp. xxv–xxviii.
10 *Ibid.*
11 *Ibid.*, Part I, p. xix.
12 This fact has been discussed with Dr. Bijoy Kumar Chaudhary, Director, K. P. Jayasawal Research Institute, Patna.
13 Jawaharlal Nehru, *op. cit.*, pp. 204–207; Amartya Sen, *op. cit.*, pp. 170–175.
14 Kripa Shankar Shukla (ed.), Part I, *op. cit.*, pp. xxvi–xxvii; Part III, p. xxiv.
15 Robert Kanigel, *The Man Who Knew Infinity*, Washington Square Press, Pocket Books, New York, 1991, p. 36.
16 Kripa Shankar Shukla (ed.), Part III, *op. cit.*, p. xxiv.
17 *Ibid.*, Part I, p. xxvii.
18 Ashok Kumar Mallik, *The Story of Number*, IISc Press-World Scientific, Singapore, 2018, pp. 99–104; Richard Courant and Herbert Robbins, *What Is Mathematics?* Oxford University Press, 1969, pp. 299–300.

19 Kripa Shankar Shukla (ed.), Part I, *op. cit.*, p. 45.
20 Alexeĭ Volkov, "Zhao Youqin and His Calculation of π", *Historia Mathematica*, Vol. 24, 1997, pp. 301–331.
21 Jan P. Hogendijk, "Al-Kashi's Determination of π, to 16 Decimals in an Old Manuscript", *Zeitschrift fur Geschichte der arabisch-islamischen Wissenschaften*, Vol. 18, 2009, pp. 73–152.
22 Frederic August Rosen (ed. and tr.), *The Algebra of Mohammed Ben Musa*, Oriental Translation Fund, London, 1831, pp. 1177–1186.
23 S. Balachandra Rao, *Indian Astronomy: An Introduction*, *op. cit.*, p. 92; Kripa Shankar Shukla (ed.), Part I, *op. cit.*, pp. 29–30.
24 Lancelot Hogben, *Mathematics for the Million*, Pocket Books, New York, 1965, p. 176; Daniel D. Benice, *Mathematics, Ideas and Applications*, Academic Press, New York, 1978, pp. 179–185.
25 Kripa Shankar Shukla (ed.), Part I, *op. cit.*, pp. xxxiv–lxviii.
26 Bina Chatterjee (ed.), *The Khandakhādyaka (An Astronomical Treatise) of Brahmagupta: Introduction, Translation and Mathematical Notes*, Vol. 1. Distributor: World Press, Calcutta, 1970, p. 73.
27 Bina Chatterjee, (ed.), *Sisyadhivriddhida Tantra of Lalla with Commentary of Mallikarjuna Suri: Part 2: Translation and Mathematical Notes*, Indian National Science Academy, New Delhi, 1981, pp. 3–4.
28 Amartya Sen, *op. cit.*, pp. xi–xii; pp. 26–30.
29 Kripa Shankar Shukla (ed.), Part I, *op. cit.*, pp. 129–130.
30 Hari K. Chandra, "Critical Evidence to Fix the Native Place of Āryabhata-I", *Current Science*, Vol. 93, No. 8, 2007, pp. 1177–1186.
31 Kripa Shankar Shukla (ed.), Part I, *op. cit.*, pp. xxxiv–lxviii.
32 Prafulla Chandra Ray, *A History of Hindu Chemistry*, Centenary Edition, Shaibya Prakashan Bibhag, Kolkata, 2002, pp. 192–196.
33 J. F. Scott, *A History of Mathematics*, Taylor & Francis, London, 1967, pp. 67–82; John Narrien, *Historical Account of the Origin and Progress of Astronomy*, Baldwin Cradcock, London, 1833, pp. 314–341; A. Pannekoek, *A History of Astronomy*, George Allen & Unwin, London, 1961, pp. 164–170.

6 Proto-historic Astronomy in India

A Review

Venkateswaran Thathamangalam Viswanathan

The origin of ancient Indian astronomy has remained one of the most seriously debated questions for more than 200 years. The *Vedāṅga Jyotiṣa*, one of the earliest surviving astronomical texts, says, "Just like the combs of peacocks and the crest jewels of serpents, so does *jyotiṣa* (astronomy) stand at the head of the auxiliaries of the *Veda*".[1] It further emphasizes that

> one learned in the Vedas, who has also learnt the lore of the movement of the moon, the sun and the stars will enjoy after death a life in the world wherein the moon, the sun and the stars move, and he will have also on the Earth an unending line of progeny.[2]

Vedas, the earliest literature of the subcontinent, mention professions, such as *gaṇaka* (calculator) and *nakṣatra-darŚa* (star-gazer), and refer to astronomy as *nakṣatra-vidyā* (star science) illustrative of proto-astronomy of that period.

Rigveda, produced between c.1500 BC and c.1000 BC, has stray references to astronomical events and calendrical implications. Of the five planets, *Rigveda* specifically mentions *Bṛhaspati* (Jupiter) and *Vena* (Venus) by name. One such Rigvedic verse mentions the moon being near nakṣatras[3] and the latter *Taittirīya Saṃhitā* provides lists of the 27 nakṣatras.[4] In its latter part, *Rigveda* suggests intercalary months to fit the lunar calendar with the tropical solar cycle. *Taittirīya Saṃhitā* was aware of the sunrise point moving northwards (*uttarāyana*) half of the year and southwards (*dakṣiṇayana*) in the next half and says that the earlier 360-day Vedic year was inadequate and suggests adding five additional days to account for the tropical year. Furthermore, it says four days are too short and six days are too long, implying a much more precise measurement of the tropical year than earlier. The Tamil literature corpus *Puṟanāṉūṟu*, dated 100 BC–AD 250, not only mentions names of a few nakṣatras but calls them the "star of the day" (*maticēr nāḻmīṉ*) associated with the 27-day sidereal cycle of the moon and the 3rd–5th century BC Tamil literature, *Patiṟṟuppattu* (59), talks about the variation of the day and night lengths.

DOI: 10.4324/9781003370642-7

Vedāṅga Jyotiṣa or *Jyotiṣavedāṅga* is a compilation of astronomical/cylindrical rules needed for Vedic rituals.[5] Principal ideas of *Vedāṅga Jyotiṣa* that perhaps were formed sometime between 600 and 400 BC have remained influential even as late as the 4th or 5th century. It prescribes a five-year luni-solar cycle (*yuga*), with 1830 civil days as its unit to bring harmony between the lunar and solar movements. It noted that 62 synodic months of 29.5 days, 67 sidereal months, and 60 solar months approximate 1,830 civil days.[6] Kātyayana *Śulvasūtra* (c. 500 BC) describes how *Sanku* (gnomon) can be used for finding the cardinal directions.[7] *ArthaŚāstra* of Kautīlya (c. 300 BC) describes diurnal and annual variations of the length of the shadow of the *Śaṅku*, and how it can be used for fixing the time of the year.

From about AD 600 onwards, we have a number of texts that show how *Śaṅku* can be used for determining the cardinal direction, local latitude, time and time of the year.[8] It is difficult to imagine that sophisticated agrarian communities of Indus culture and megaliths, pre-dating Vedic period, would have flourished for thousands of years without even some form of a calendar for civil and religious use and knowledge of crude naked-eye astronomy when pre-historic and proto-historic cultures around the world sport some form of archaeo-astronomical and ethno-astronomical traditions.[9] However, we know very little about the Palaeolithic, Neolithic, Indus and Iron Age cultures that thrived in South Asia before the advent of the historic period. This chapter looks at the contemporary research in this direction for ferreting out and documenting the probable astronomical knowledge and its application during the Neolithic, Indus and Megalithic proto-historic periods in South Asia.

Applying the tools and methods of archaeo-astronomy and ethno-astronomy, recent researchers have unearthed probable astronomical traditions dating back to the proto-historic period. In order to have a proper understanding of the topic discussed so far, the present chapter is divided into five different sections:

- "Proto-history of India" section presents a snapshot of proto-historic South Asia.
- "Probable Astronomy of Proto-historic Period" section introduces the basics of positional astronomy.
- "Glimpse of Neolithic Astronomy" section looks at the possible evolution of India specific astronomical concepts, such as *tithi* and *nakṣatra*, from the Neolithic calendrical practices.
- "Indus Valley Civilization" section provides a brief review of the astronomical elements teased out of Indus excavations and research into its script.
- "Megaliths" section projects a summary of the exciting contemporary work on the astronomical significance of megaliths in India.

Proto-history of India

South Asia is a political fabrication which includes the modern nations of Bangladesh, Bhutan, India, Nepal, Pakistan, Sri Lanka and the Maldives. This region has had linguistic, historical and archaeological links with parts of Afghanistan and Iran. The Indian subcontinent, a geographical unit, refers to the landmass below the Himalayan range distinct from the Eurasian continent. In historical times, this geographical unit has been loosely associated with 'India'. Persian traders, around the 6th century BC, referred to people living in this region, beyond river Indus, by the term 'Hindus', a corruption of Indic Sindhu. The Greeks, led by Alexander the Great, invaded this region in the 4th century BC and called the river Sindhu in Greek '*Indos*' and India to refer to the land beyond the river. Today, South Asia, 4.4 million square kilometres in extent, is home to about 1.5 billion people, i.e. one in five of the world's population resides in the subcontinent. About 657 languages are spoken in this region that belong to four main language families of the world: Indo-Aryan, Dravidian, Austro-Asiatic and Tibeto-Burmese.

As most of the pre-historic period is devoid of contemporary written records, primarily the evidences, gathered through archaeology, inform and shape our interpretation of the past. The same may be supplemented with anthropological and ethno-graphical myths and folklore to get a rough picture of the life and time of the pre-historic age. Excavations at Upper Siwaliks, the Acheulean quarry of Isampur, Palaeolithic Madrasian handaxe culture in South India, the Palaeolithic and Mesolithic rock shelters of Bhimbetka, the late Pleistocene caves of Kurnool, the Mesolithic cemeteries in the Ganga Plains, the early Neolithic settlement of Mehrgarh, the Harappan cities in the Indus Valley, the Iron Age centres in the Ganga Valley and recent excavations in Kodumanal and Kodungallur provide us with a fascinating glimpse of the pre-historic life in the Indian subcontinent. These archaeological finds are supplemented with studies on cultural anthropology, geology and climatic modelling studies to shed further light and assist interpenetration of the available data. A number of comprehensive works documenting the remarkable proto-history and pre-history of the subcontinent have been published.[10] Recently, genetic studies have emerged as a powerful persuasive tool in validating certain crucial issues concerning the pre-history of India.

Nevertheless, despite significant field work and research, serious hurdles persist in the fuller understanding of human evolution and societal change in India and drawing implications from this for synthesizing comprehensive world history. In addition to the above, scholarly investigations are hampered by narrow nationalism and political agendas, resulting in serious distortion.[11]

Evolution of Modern Human Population

About six million years ago a new species of *hominin* with bipedal locomotion evolved. Since then, many species of *hominins* have evolved and some of them coexisted for about 3.5 million years. Until recent times, the multi-regional hypothesis is that *Homo sapiens sapiens* (modern humans) evolved from *Homo erectus* at multiple locations around the world, resulting in the visible 'racial difference' of the modern human population. However, genetic studies persuasively show that all the living *Homo sapiens sapiens* around the world, irrespective of race or geography, share a common African ancestor who lived about 200,000–150,000 BP (before present).

Arrival of Modern Humans to the Indian Subcontinent

How modern humans came to populate the Indian subcontinent is one of the fiercely contested domains. Recent studies indicate the southern coastal human dispersal route, in addition to the commonly hypothesized northern-over-land route of modern human dispersal populating South Asia. The first humans, dated to 50,000 BP, landed on the Southern Indian shores and populated Andaman and Nicobar Islands and continued towards Indonesia and Australia. The second and recent migration, it is posited, was through traditional West Asia around 30,000 PB, through the northwest region.

The only known pre-modern *hominin* fossil, in the subcontinent of late Acheulean or early Middle Palaeolithic, has been recovered from Hathnora in the central Narmada Valley, dated to about 2.5 million years BP. The earliest modern human (*Homo sapiens*) skeletal remains in South Asia come from the sites of Fa Hien and Batadomba Lena in Sri Lanka which yield dates of 30,000 and 28,000, respectively, for human remains associated with microlithic tool assemblages. Nevertheless, 45,000 BP, upper Palaeolithic assemblages at Site 55 in Pakistan and 26,210 BP at the 16R dune in Rajasthan indicate clear evidence for modern human behaviour in the Indian subcontinent. Microlithic sites have been found at Bagor in Rajasthan, Langhnaj in Gujarat, Sarai Nahar Rai, Mahadaha and Damdama in the Gangetic plains and Adamgarh, Bhimbetka and Ghagharia in Central India. Almost all of these sites are less than 30,000 years old, and they indicate a definite advanced technology than the Palaeolithic.

Mesolithic India

The transition from the Palaeolithic to the Mesolithic witnessed the emergence of a new type of stone tool known as the microlithic. The microlithic was functionally a more useful tool than a large stone pebble because it could be used to shape many other tools, for instance to make knives or scissors. Although a direct link between microlithic tool assemblages and modern humans is poorly established, some microlithic settlements are clearly associated with *Homo sapiens* which have led archaeologists to treat all microlithic sites as the Mesolithic culture of modern humans. The Mesolithic period in South Asia lasted until about 3800 BP, well into the copper age.

Emergence of Neolithic Culture

With the gradual transition to Neolithic culture, during the late Pleistocene era nomadic, hunting-gathering way of life gave way to a more sedentary, settled and food-producing way of life. The climate was getting warmer; the Pleistocene ice cover was retreating, and as the Holocene epoch approached, evidence of the Neolithic Revolution could be seen across India too, as in the case of Egypt, Mesopotamia, Mesoamerica and China. The stone tools, such as polished ground-stone axes, stone sickles, hammer stones and stone blades, are seen in these settlements, and these tools are also far more varied and efficient than those of the earlier periods. Domestication of plants and animals by humans began during this epoch. The evidence from various Neolithic sites spread across the subcontinent suggests that farming, pastoralism and the establishment of settlements did not all emerge simultaneously, and at all sites. The earliest evidence of agriculture in South Asia dates back to 13,000 BP at Mehrgarh which provides the earliest evidence for plant and animal domestication and village life. Other Neolithic transitions occur over the subsequent millennia in diverse regions of the subcontinent with a range from 9000 BP to 3000 BP.

The invention (discovery or development) of agriculture and polished stone axe, which are hallmarks of this epoch, brought about dramatic changes in the economy, technology and demography of human societies. Stone axe helped the clearing of forests to create farm lands and agriculture implied a steeled way of life. The introduction of agriculture saw the shifting of habitat to river flood plains which provided rich alluvial soil and a perennial supply of water for farming activity. Farming activity is suspected to be an important motivation for the study of seasons and thereby aids the emergence of proto-astronomy.

Probable Astronomy of Proto-historic Period: Need to Study Stars

On a given clear moonless night, about 6000 stars are visible to the naked eye, and some of them exhibit striking patterns. Asterisms, such as Big Dipper, Orion, Pleiades, Leo and Scorpio, are readily recognizable, and various cultures around the world have identified them from pre-history. The absence of multi-wavelength telescope, photograph, spectrography and very precise time measuring devices puts severe limitations on the pre-historic astronomy. It is, but, evident that only those cyclic changes that could be perceived, discerned (with the naked eye) and measured (with the aid of simple instruments) could have been studied in the proto-historic period. As Giulio Magli observes,

> Many ancient cultures acquired a profound knowledge of ... celestial bodies available for their observations were only those visible with the

> naked human eye ... The continual changes in the sky observed by them were, therefore, for the most part, due to apparent movements. In spite of this, many ancient astronomers devised ingenious methods of observation and made highly accurate measurements, which were reflected and exploited in the projects of the monumental architecture of their historical epochs.[12]

Archaeo-astronomy is an amalgam of the history of astronomy, astro-archaeology and ethno-astronomy to enable us to reasonably infer from the available evidence the astronomy present and practised in the pre-historic and proto-historic periods. Krupp[13] broadly defines archaeo-astronomy as

> calendrics; practical observation; sky lore and celestial myth; symbolic representation of celestial objects, concepts, and events; astronomical orientation of tombs, temples, shrines, and urban centres; symbolic displays involving celestial phenomena in the natural environment; traditional cosmology; and ceremonial application of astronomical tradition.

As Gurrukal and Varier[14] have pointed out that reconstructing belief systems and ideology from material remains seems daunting and are always exercises bringing into play the archaeological imagination. Astro-archaeology attempts to retrieve astronomical information from the study of the orientation and layout of the architecture. The history of astronomy is largely dependent on written records which are scant in pre-historic and proto-historic periods. However "Astroarchaeology is a field methodology, often ... retrieving astronomical information from the study of alignments associated with ancient architecture and the landscape",[15] while "Ethnoastronomy is a branch of cultural anthropology that draws its evidence from the ethno-historical record and ethno-graphic studies, and attempts to answer 'why' astronomy, whereas the former studies deal primarily with 'how' astronomy".[16] Inferred in this light, various essays in recent collections such as Ruggles (2015)[17] along with Thurston (1996),[18] Evans (1998)[19] and Aveni (1989)[20] provide examples of calendrical astronomy during pre-history and proto-history.

Motivation for Studying Stars

As awe and simple curiosity of the human mind could have turned them towards the heavens, therefore, a much deeper human need is behind the impulse for early civilizations across to look at the sky and meticulously study the pattern of apparent motion of celestial objects. Civilization and culture imply some form of social organization. Social organization demands some form of ordering of time and space. Cultural anthropological study of hunter-gathering and subsistence farming tribes who are still beyond the reach of 'civilization' clearly shows that some form of calendar are prevalent and their conception of day, month, year and any unit for longer duration (such as '*yuga*' in Indian astronomy) shape their collective activities and synchronize repetitive sequences of events in their social life. Activities such as planting, harvesting, hunting and recollecting (say communal gathering), and other economic, cultural and socio-religious rituals are ordered not only in space (where they can/should take place) but also in temporal domain (when they should take place).[21]

It is fundamentally through calendars that cultures acquire temporal framework, synchronize events, arrange them chronologically and determine time intervals between events. Most calendar systems are based on the movements of heavenly bodies, as these astronomical phenomena are regular and periodic and, therefore, offer the possibility of measuring physical time. Indeed other events, such as the flowering of certain trees, are also taken as a time marker in many cultures, but often they are region specific and, hence, remain a practice only within a small group dwelling in a restricted space. As societies acquire some level of sophistication in their culture, trade and network with other communities get strengthened, and invariably they fashion the calendar around periodic movements of astronomical objects to ensure regular, predicable and consistent temporal ordering over a larger geographical area.

Nilsson[22] observes that in addition to astronomical events, such as a heliacal rise of a specific star, a sunset behind an important landmark and the reappearance of the first moon, primitive cultures also use recurrent environmental events, such as the first rainfall, the arrival of certain birds and the flowering of specific plants, as temporal markers. For example, in Indian folklore, the first appearance of the migratory bird Pied Cuckoo (*Clamator jacobinus*) is described as a harbinger of the monsoon. Although these environmental cues are unevenly distributed and do not provide uniform units of time, they are culturally defined and socially meaningful 'guide marks' informing people when a given activity should start. As the civilizations grow and the number of habitations across the region integrate resulting in some form of proto-state formation, standardization and uniformity throughout the region are imperative. Perhaps, astronomically based calendars emerged as cultural devices used to synchronize distinct social, cultural and natural phenomena for a more cohesive social formation.

Typically, in any calendar, the measurement of time larger than a day is linked to one or more astronomical cycles. The phase cycle of the moon, seasonal cycle of appearances and disappearances of stars and asterisms, and movement of the position of sunrise or sunset at the horizon are some of the cycles that are used in framing the basis of a calendar. Even exotic natural cycles, such as the cycle of appearance and disappearance of Venus in the sky, have been, for example, used in the Mayan calendar. "Cultural diversity, together with the fact that the principal astronomical cycles do not fit neatly together, has led different communities to create an extraordinary range of calendars fitted to particular situations and social needs, often showing remarkable ingenuity".[23] We come to know of a description related to a distinct cultural specific calendar.[24] We also come across a description of diversity in calendars around the around.[25]

Cosmogony

Space and time seem to be so fundamental and familiar to our daily experience that we hardly ponder about it. From a scientist's perception, time and space appears to be an objective entity which may be so. However, in so far as human societies are concerned, concepts of space and time are cultural products, yet remain embodied in physical objects, events and processes of the objective world. Furthermore, making sense of the world involves recognizing correlations between changes taking place in the natural world and using these observed events to predict what is yet to come. People around the world create rules by which the world works and locate their place in it and fabricate a concept of cosmogony for their culture. Not only the objective impact of the cosmos but also the culturally imagined view of it affects the ways in which people interact with their physical, social and cultural environments. Thus, invariably, cosmogony is embedded in "mythological and ritual structures and in everyday practices and taken-for-granted values, and embodied in symbolic representations, landscape features, ceremonies and other cultural products and organize and structures societies".[26] As fabricated cosmogony has to be consistent with the observed world around us, human civilizations would require knowledge at least to a certain degree of basic astronomy.

Measuring Time: What Is a Day?

Today, one defines the one rotation of the earth around its axis as a basic unit of time, the day. However, one should hasten to add that even up to AD 1200, the idea that the earth is spherical and rotating about itself was a radical idea, and even some astronomers scoffed at it until the recent period. For example, we may note a recount of how even the famous Brāhmagupta vehemently attacked and refuted the diurnal rotation of the earth.[27] The revolution of the moon around the earth gives rise to the concept of a month and the revolution of the earth around the sun, a year. Given that the ancient world was devoid of sophisticated astronomical instruments available today, the model of the cosmos that prevailed was largely anthropocentric. From anthropocentric perception, what we perceive is a stationary earth with heavenly bodies revolving around it. Thus the 'real' motions like the rotation of the earth around its axis, or the revolution of the earth around the sun was perceived as the diurnal motion of the sun across the sky and as the annual revolution of the celestial sphere. Further, they had to work out a way of measuring these motions.

A vertical pole can tell us when the sun is crossing the zenith (overhead). At the local noon, when the sun is at its zenith, the shadow cast by it would be the shortest for the day. Thus, with a simple gnomon (*Śaṅku* in Indian astronomical tradition), one can measure two successive local noons. The time interval between them gives us a measure of the civil day (~24 h). Alternatively, one can also measure the time taken for a specific star to come to the zenith on two successive days, and this gives rise to a concept called sidereal day (23 h 56 m 4.09 s). Sidereal in Latin means 'star' and measurements with respect to stars are called 'sidereal' periods. If mathematical astronomy grows to a certain level of sophistication, one can look at the time period between the vernal equinox, say crossing the zenith on two successive days. This is called the tropical day and the duration is the same as the sidereal day.

One Month

Similarly, the period between the same lunar phases, say full moon to next full moon, is called the synodic month (the length of the mean synodic month is 29.530589 days), while the interval between the two consecutive conjunctions between the moon and a given asterism (as seen from the earth) is called the sidereal month (27.321661 days). The Greek word 'synodic' means 'meeting' or 'assembly'. It is motion relative to the conjunction or alignment of two heavenly bodies. Both the cyclic lunation and the moon appearing near a star are easily noticeable to even the unaided eye. Thus, they too can be measured without sophisticated aids.

Tropical Year

As the earth goes around the sun, the time it takes to return to the same place is one sidereal year. However, if we consider the return of the earth to one of the two intersection points of the earth's orbit around the sun with that of earth's equatorial plane, then we get a tropical year. In other words, the time interval between two successive vernal equinoxes is the tropical year (365.24219 days).

But how does this play out in anthropocentric perception? The axis of the earth's rotation is tilted at 23.5° with respect to its axis of revolution around the sun. The direction of the axis of the earth's rotation is always pointed towards the North Pole star, as the earth goes around the sun. From a geocentric perspective, this would be equivalent to saying that the ecliptic apparent sun's path in the celestial sphere is at the angle to the celestial equator by the same 23.5° angle. The declination of the sun is the measurement of the angle between the sun's rays and the earth's equatorial plane, in other words the angle between the position of the sun on the ecliptic and celestial equator.

The sun's declination varies throughout the year. During a course of a year for a point in northern latitude, the declination of the sun goes on increasing every day and reaches the maximum declination angle of 23.5°. Thereafter, the declination reduces every day and drops to a minimum at negative 23.5° declination. This would imply that the sun's declination during the year varies by about 47°. The days on which the sun's declination is zero are called equinoxes. That is, the sun is exactly on one of the two points of intersection of the celestial equator and ecliptic. At the equinoxes, at all the places on the earth, the nights and days are of equal duration (i.e. 12 hours), hence the term equinoxes (equal nights from Latin '*aequinoctium*'). In the current epoch, the autumnal equinox occurs on about 22 September when the sun crosses the celestial equator as it moves southwards, and the vernal equinox occurs on about 21 March when the sun crosses the celestial equator as it moves northwards. The day on which the sun's declination is maximum, that is +23.5°, is called the summer solstice, and when it is –23.5°, it is called the winter solstice.

When the declination of the sun reaches its upper or lower limit, the solar declination of the sun seems to stop before it changes direction. As of this day, the sun appears to 'stand still' without change in the declination. Derived from Latin '*solstitium*', these days are called solstice. In the current epoch, the summer solstice (mid-summer in the Northern Hemisphere) occurs on about 21 June when the sun's declination reaches 23.5° north (the Tropic of Cancer), and the winter solstice (mid-winter in the Northern Hemisphere) occurs on about 21 December when the sun's declination is 23.5° south (the Tropic of Capricorn).

How do we perceive the change in the declination of the sun? The changing declination implies that the sunrise point would vary from day to day. It is customary to measure the angle of a point at the horizon from the north, clockwise. This angle measure is called the azimuth. This varying declination of the sun has a striking and dramatic effect on the earth. The variation in the sun's declination through a year affects its apparent rising (and setting) point on the horizon. Depending upon the observer's latitude and the declination on any given day, the azimuth of the rise (or set) of the sun changes. In the Northern Hemisphere, beginning at the winter solstice, the azimuth of the sunrise point decreases compared to the previous day until it reaches the summer solstice. At that mid-summer date, it has the smallest azimuth of rise (i.e. most northern). It stops decreasing and, thereafter, the sunrise point is more to the south compared to the previous day until the winter solstice is reached again. During this period, the azimuth of the sunrise point keeps on increasing every day until the sunrise point reaches the maximum on the winter solstice.

In other words, from a given location, the sunrise point would appear to move either in a northerly direction or in a southern direction. In Indian astronomical tradition, the winter solstice, marked as the commencement of *uttarāyana*, is the northward movement of the sunrise point on the horizon, and the summer solstice was called *dakṣīṇāyana* which is the southern movement of the sunrise point.[28] Thus, one complete oscillation of the sunrise point provides a tropical year.

Sidereal Year

The time interval between two consecutive conjunctions between the sun and a given asterism, as seen from the earth, defines the sidereal year (365.256360 days). Noticing the heliacal rising of the stars provides an easy way to measure the sidereal year. A heliacal rising of a star is defined as its first appearance above the eastern horizon just before sunrise. Likewise, the heliacal setting is defined as the last day a star or a planet is barely visible after sunset in the evening twilight. As it becomes increasingly proximate and eventually in the glare of the sun, the stellar object becomes invisible. Thereafter, as its position is in the same direction as the sun, it rises and sets with it. Consequently, it remains hidden from visibility both during the day and night. However, as the sun moves along the ecliptic, the angle between the star and the sun increases. When the separation is between 8° and 20° of zodiacal longitude, the star begins to emerge, briefly, immediately before sunrise; its first appearance during the twilight is known as its heliacal rising.

Ancient cultures gave particular emphasis to the heliacal rising and setting of the stars since these could be used as reliable indicators of agricultural conditions.

> Egyptians put a great deal of emphasis on the heliacal rising of the star Sirius (which they called '*sopdet*') that was, for a long period of time, associated with the annual flooding of the Nile upon which Egyptian agriculture depended.[29]

Heliacal observations do not call for any special skill or advanced astronomical theories.[30] Many cultures across the world, such as from Egypt, Mesopotamia and elsewhere, have taken note of the heliacal rising and setting of the stars. Heliacal observations, to ascertain the turning points of the seasons, appear to be the most common practice in both the literate and pre-literate cultures. Leaman[31] et al. found the associations between heliacal or acronychal rising the prominent stars of the celestial birds, animals and the lifestyles of their terrestrial counterparts. Parpola[32] claims, "The heliacal rise of Aldebaran at vernal equinox marked the new year".

Due to the precession of the equinoxes, the sidereal year is about 20 minutes longer than the tropical year. The tropical year is the interval at which seasons repeat and is the basis for the calendar year. Thus, there are many possible understandings of days, months and years.

Cosmic Wheel

While in the day, the sun seems to rise in the east and set in the west with the sky seeming to be stationary, at night the whole starry hemisphere appears to move from the east to the west.

Although the perception of the spinning sky is the same all over the world, there are significant differences. At the equator, the north and south celestial poles lie on the horizon, and all the stars appear to rise at the right angles to the horizon and move in a semi-circle across the sky. However, for people at a higher latitude in the Northern Hemisphere, some stars and asterisms, called circumpolar stars, neither rise nor set but appear to move in a circle around the Pole star. In a similar manner, in the Southern Hemisphere, some stars appear to go around the south celestial pole which appears at an angle above the northern horizon. Other than at the equator, at all other places on the earth, diurnal arcs of stars make acute angles with respect to the horizon. At the North and South Poles, the angle is zero, and, hence, they appear to move in a circle around the celestial poles. Thus, careful observation of the motions of the night sky readily leads us to the notion of a celestial sphere – a sphere bedazzled with stars enveloping the earth.

While just as the stars, the sun, the moon and the planets share diurnal westward movement; careful observations readily show that the movements of these seven bodies are quite distinct from the stars. Except through telescopic observations, the relative motion of the stars cannot be discerned. Thus, stars appear to be fixed on the celestial sphere, thus the nomenclature 'fixed stars'. Just as the planets move among the 'fixed stars', birds chirp back and forth among the herd of goats says the 2nd–3rd century BC *Paṭṭiṉappālai* (67–77), indicating the identification of 'planets' distinct from the 'stars' from the motions that they exhibit.

However, the sun, the moon and the planets exhibit relative motion with respect to 'fixed stars'. In other words, if a star rises (or sets) with the sun today, it would not be the same the next day. Thus, the heliacal rising (or setting) of the stars could also be used to keep track of the yearly solar cycle. The notion that the heavens constitute a great sphere surrounding the observer is an ancient one. In ancient cosmogony, the rotating sky was compared to the potter's wheel. A description of the deployment of analogy to a potter's wheel may be explained in the Hindu cosmography.[33] The Pole star was thought to be the axis, and the rest of the stars were thought to be tied to it by an invisible rope. Cosmic wind drives the stars as they move around the earth.[34]

In the background of the 'fixed' stars-studded celestial sphere the sun appears to be displaced eastward by 1° per day. Consequent to this apparent displacement, the stars appear to rise approximately four minutes earlier each evening and set four minutes later each morning. Further, except for two instants during the year, the sun's annual motion is not strictly parallel to the celestial equator.

The moon appears to be displaced approximately 13° eastward relative to the 'fixed' stars on the celestial sphere per day. Therefore, the moon rises and sets almost 45–50 minutes earlier each day. The moon's orbit in the celestial sphere tilted at about 5° to the ecliptic. The sun appears to complete one revolution around the ecliptic once a year, while the moon does it in about 27.3 days, which is called a sidereal period. However, as the earth is revolving around the sun, the displacement of the earth along its orbital path causes the time that it takes for the sun, earth and moon to return to the same starting alignment, that is the completion of one cycle of lunar phases, is slightly longer than the sidereal period, takes about 29.5 days, which loosely is approximated to 30 days of a lunar month. As the moon revolves around the earth, its declination changes in the course of a month, and consequently the moon's rise and set points on the horizon vary much more rapidly than that of the sun.

Heavenly Objects

In the ocean of star-studded sky, seven objects, the sun, moon and five visible planets (Mercury, Venus, Mars, Jupiter and Saturn), stand out due to their peculiar motions. These 'heavenly' objects were known to most of the civilization since antiquity. The diurnal motions of planets are slower than the sun and moon but are noticeably faster than those of 'fixed stars'. Thus, although planets too appear like a big and brighter star, they stand out in the ocean of brilliant starry sky. Nevertheless, the apparent motions of planets are more complex, sometimes halting their eastward motions (prograde) and briefly moving westward (retrograde) before resuming their usual eastward motion.

Seasons

A crucial consequence of the large annual variation in declination is the profound variation in the flux of solar energy received at a given point on the earth during the course of a year. This variation in the receipt of solar flux is perceived as seasons. The Northern Hemisphere experiences the longest daylight during the summer solstice. At Poles, for six months, it is a long day time, with the sun never going below the horizon. For the next six months, the sun does not appear above the horizon, resulting in a long night. For instance, when the declination of the sun is 23.5° during the summer solstice, north of the Arctic Circle receives sunlight 24 hours a day, while Antarctica receives 24 hours of darkness.

It is obvious that the Northern Hemisphere and Southern Hemisphere will have opposition seasons. When the Northern Hemisphere is tilted towards the sun, the summer season occurs while the Southern Hemisphere experiences winter. Then after a period of six months, the Northern Hemisphere undergoes winter while it is summer in the Southern Hemisphere. Furthermore, variation in the declination of the sun has effects on its own altitude and the duration of daylight. The time interval of the sun above the horizon that is the day time changes with the seasonal change in declination. During the summers, the sun is above the horizon longer than in winter at all latitudes except the equator.

As the changing declination of the sun causes the seasons, it was important to measure it for calendrical purposes. One can do it in two ways. First, a crude method is to measure the azimuth variation of the sunrise point. In absence of microscopic angle measuring devices, the precision measurement is difficult, as well for about ten days the sunrise point would appear 'standing still' at the solstice points. On the other hand, with a gnomon of sufficient height, it is easy to mark the shadow length at local noon time much more precisely. In the Northern Hemisphere, the noontime shadow on the summer solstice would be the shortest in the year and will be longest on the winter solstice. Furthermore, the time of the year from the length of the noon shadow.

Ecliptic

During the year, the moon and the planets appear to move in a restricted region in the celestial sphere close to the ecliptic. This band extending about 8° from each side of the ecliptic is called the zodiac. This band was divided into 12 equal regions, called 'house', and traditional zodiacal constellation patterns were imagined out of the stars in each 'house'. While the daily rhythm of the day and night and the monthly cycle of lunar phases are readily apparent to all, the same pattern of stars (asterisms) reappearing overhead exactly at 12 in the midnight after completion of about 365 days at a given location requires years of observation, collection and recording of data from various locations and to abstract this pattern. This also presupposes some level of social organization and specific arrangements for careful observation of the dynamic sky. Thus, any civilization that exhibits an understanding of year, in particular, sidereal year, would have achieved some level of sophistication in astronomy.

Glimpse of Neolithic Astronomy

Certain ideas of Indian astronomical traditions appear to date back to antiquity. We see them in both the Vedic literature, the Dravidian corpus and tribal cultures. Recognition of the visible planets by their characteristics, colours, the origin of the concept of *nakṣatra* and *tithi* in the calendrical system and the concept of the Pole star being the centre of a giant wheel to which the sun, moon and stars are tied through invisible aerial cords are some such ideas. In accounting for the historical evolution of these concepts of ancient Indian astronomy and cosmology, many "scholars have fought shy of the idea of assigning them a native Indian origin"[35] and have speculated Mesopotamian influence in the proto-historic period. In reaction, some scholars have gone 'native' and rejected any suggestion of influence from outside in knowledge production. Ideas have always moved, spread, taken in and modified by the receiver. Knowledge has grown only because of this process of constant interaction between cultures and schools of thought.

Tithi and Nakṣatras

Ancient Indian astronomy seems to have concentrated on lunar observation creating nakṣatras to the exclusion of zodiacal signs. The idea of a zodiacal sign (*RāŚi*) has been incorporated at a later date. These nakṣatras, lunar mansions, are unique to India, although a somewhat similar concept appears in Chinese Civilization. *Tithi*, lunar day, being equal to 1/30th of the synodic month, and 61/62nd of a *sāvana* (natural) day, is another unique element in Indian astronomy. These two derived units of time have been given importance in the Indian calendrical system and are in use even today for ritual purposes.

Each of the 30 tithis within a lunar month was given its own name, and a *sāvana* day was named after a particular *tithi* which began at sunrise. Similarly, each of the 27 nakṣatras, corresponding to 1/27th of one sidereal month (~27.3 days), has an asterism associated with it. Each *sāvana* day is also associated with a particular asterism (nakṣatras) in which the moon appears at the time of sunrise. Some scholars have postulated that Babylonian was passed on to Hellenistic astrology which reached the shores of India when contact with the Greek world was established. However, there are a number of suggestive evidences that prompt us to interpret indigenous development of the idea, perhaps going back to Palaeolithic times.

Calendar Sticks of Nicobar

The diurnal motion of the sun in the sky is the most conspicuous natural marker of the progression of time. The cycle of sunrise, sunset and again the sun rising the next day constitutes a civil (*sāvana*) day. Cyclic lunations of the moon are another readily apparent natural periodic movement that helps mark the measurement of a larger unit of time. *Rigveda* describes the moon as '*māsa kṛt*', the maker of the month.[36] The Stone Age civilizations would have found cyclic lunation of the moon handy to mark events that have monthly periodicity and also for temporal organization of social life.

World over, several artefacts dating back to the Palaeolithic and Megalithic periods, collectively identified as 'calendar sticks' with carved tally markings, have been uncovered. Although other archaeologists dismissed them as tally mark records of hunting, Alexander Marshack[37] convincingly interpreted many of these engraved lines as lunar calendrical notations dating back to the upper Palaeolithic period. These tally marks found on bones, wood and at times etched on clay have been objects of investigation by archaeologists for many years. From Native American sun sticks, Mayan wooden calendars to calendar sticks of Siberian hunters, the practice of recording the lunation by the carving of notches appears to be common throughout the world.[38] A survey of the literature by Hayden, Brian and Suzanne Villeneuve[39] shows that "detailed solstice observances were common among complex hunter-gatherers, often associated with the keeping of calendars and the scheduling of major ceremonies". An analytical approach has also been presented by F. d'Errico[40] and Marshack[41] with regard to the views mentioned above.

Not surprisingly, two wooden sticks that belong to the upper Palaeolithic period (11,000–8000 years BP), described by Marshack, appear to be one of the earliest materials of evidence for astronomical activity in the Indian subcontinent. From his Nicobarese informants, Marshack identified that the two calendar sticks collected by him were indeed being actually used as lunar tallies. A number of colonial administrators also vouch for the prevalence of the practice of carving notches in wood and creating calendar sticks among the Nicobarese. Temple observes,[42] "The Nicobarese keep rough calendars by notches on wood. They reckon time by the monsoon seasons or period of regular winds". However, a recent survey indicates[43] that "people of Nicobar only vaguely recalled such a method but were no longer aware of the process".

Made from soft tropical white wood, the Nicobar calendar sticks are shaped to appear like a knife or scimitar. Notch marks appear on the edge and also on the flat surface.

> The notches are meant to go as 10–6, 10–4 = 30 showing the lunar phasing as from new (or the last sighting of the crescent) to quarter moon as 10 days and quarter to full as 6 days; full to next quarter another 10 and from the quarter to the new moon or the last sighting of the crescent before new moon as 4 days.[44]

However, all tally clusters do not total to 30; some are 26 and a few even 31. Bad weather or cloud cover could have prevented the record keeper from observing and noting the lunation. Set of tally marks are clustered together with chevron marks indicating, perhaps, the completion of a cycle that is the passage of a 'month'. When all the space is used up, further months are engraved across earlier ones, resulting in a cross-hatched pattern.[45] On the other hand, Brumm[46] says, "It would be difficult to argue that any of them could have functioned as portable devices for storing calendrical, lunar or other information, as has been inferred for at least some prehistoric engraved objects" about pre-historic engraved artefacts, adding a word of caution.

Engraved Markings

Wakankar and Brook[47] noted similar engraved ostrich egg shells excavated from an open campsite, upper Palaeolithic layer near Patna. Wakanker also reported finding a bone engraved with cross hatching in an upper Palaeolithic layer and another in a Mesolithic layer at Bhimbetka.[48] The pre-Harappan site of Kot Dijian excavations, 40 km east of Mohenjodaro and dated to 2600 BC, showed a pot painted with a figure that appears to be a horned human head. The ends of the horns almost touch each other and are decorated with white paint. The marking appears to suggest phases of the moon "from new moon to full moon at the centre and back to new moon to complete the cycle".[49]A similar horned head gear is present on the famous Mohenjodaro seal with markings on the horns suggestive of the lunation. A similar horned deity is seen on a pot in the Neolithic site of Burzahom.

Chowra Lunar Calendrical System

This part is summarized from Kumar and Varadarajan.[50] Perhaps daily recording of the moon's lunation was not an idle pastime but intimately connected with the economic activity of the Nicobarese-Chowra fisher community. The natural resources of Chowra are restricted, and the islanders had to necessarily sail to the other islands to garner the required resources for survival. Curiously, Chowra lunar calendrical system has specific names for each of the lunations, alluding to the concept of tithis in the Siddhanthic astronomical tradition. For example, the 'moonless night' (new moon) is called *iliuo si*, and the day before the new moon, the last waxing moon, is called *ngngo hing*. The day after the new moon, called *kanat*, is visible above the water in the west. The sailing practices of Chowra were based on stellar reckoning. To escape from the scorching sun, they preferred to sail at night. They used the position of stars in the night sky and the rising and setting point in the east and west horizon to identify the directions in the open sea. For example, '*kui tha rulu*', a bright star in the northeast, was the star indicator for Car Nicobar. When sailing from Chowra to Car Nicobar, between 2 am and 3 am, they set their course following the lower straight line of the Big Dipper, '*ngeung nga*'.

The lunations are associated with the dynamics of the sea around Nicobar, and the moon and its shape in the sky give an easy indication of the condition of the ocean currents, winds and tides. Quotidian variations in meteorological conditions are found enshrined in the lunar calendrical system of Chowra. The corpus was implicit rather than explicit. For example, on days 15 and 16, counting from the new moon day, the lunation would be a waxing half-moon called *alhinot* and there would be no current in the sea. It was dangerous to sail when *lon panlechö* (west to east current) coincided with high tide and *lon kantuko* (east to west) with low tide. Experts in weather conditions and seasonal change were called *tamol sahiöh*. This functionary set the date for all seasonal festivals and his advice was taken for all important sea-going activities.

Rock Art

The rock arts are anything but direct. They are complex images and symbols that make it difficult to interpret them unambiguously. However, the cosmic symbolism and images of the sun, moon and stars are not hard to miss or mistake. Rock art from Palaeolithic and Neolithic times has been identified all around the world. A number of scholars have attributed a variety of astronomical functions to rock art which include representation of specific historical celestial events (such as the Crab nebula supernova explosion), symbolic representation of cosmic myths, star maps and shaman shrines imbued with cosmo-magical power as part of archaeo-astronomical rock art studies.

Rock art Onake Kindi hill in Karnataka that depicts a "complex representation of a human burial with astronomical images of the sun, moon (including its phases)",[51] and other complex images suggest, perhaps, a relationship between a human burial and the cosmos which, in turn, hint at an intellectual philosophical presupposition and the complex relationship between myth and astronomy in the early days of human civilization. A rock art image found at Chillas in Kashmir is interpreted to show 'Sun God' with a halo behind him.[52] The division of the halo disc into four quadrants is interpreted as indicative of the four seasons. Sopore stone carving is interpreted to be a depiction of a meteor shower in the upper Palaeolithic period.[53]

A stone engraving, pre-dating 2000 BC found in Burzahom near Srinagar, is believed to be an early observation of a supernova or perhaps a comet.[54] The rock engraving has two suns in the sky and has the sun and moon so close to each other with comparable brightness. A supernova HG9, dated around 5000 BCE, that satisfied all the criteria like having a magnitude comparable to that of the sun or moon and a location close to the ecliptic is depicted in the rock engraving. Other features of the rock art, such as hunter, stag and dog, also fitted well with the Orion, Taurus, Andromeda and the Arietis-Cetus region where the HG9 supernova is located. Thus, a strong case is made to interpret this rock art as the recording of a spectacular transient event. Iqbal et al.[55] opines, "Collectively, these Kashmir sites appear to depict meteorite impacts, a supernova, various constellations and possibly the changing of the seasons".

Indus Valley Civilization

An excavation campaign by Sir John Marshall in the 1920s brought to light the ruins of a large brick-laid city near Harappa, presently situated in Pakistan. Astonished at the advanced civilization that the excavations hinted at, archaeologists rushed to excavate many more potential sites in the region revealing hitherto unknown civilizations.

Early excavations were along the river Indus, giving the civilization the name 'Indus Valley Civilization'. Since then, extensive excavations in India, Pakistan and Afghanistan have unearthed extensive Indus Civilization sites much beyond the Indus plains. Until now, about 1400 Indus sites, of which 917 are in India, 481 in Pakistan and one in Afghanistan have been excavated.[56] Sites, dating far back to 7000 BC to as late as 1300 BC, have been found; however, the peak mature period is estimated to be around

2600–1800 BC. During this mature period, historians postulate that there were more than 5000 habitations spread over an area of more than a million square kilometres. With about 120–340 persons per hectare, the largest of the Harappan civilization, Mohenjodaro and Harappa had a population of about 85,000 and 65,000, respectively.[57]

Astronomical Implications in Indus Script

More than "4,000 specimens of this script survive on stamp seals carved in stone, on moulded terracotta and faience amulets, on fragments of pottery and on a few other types of inscribed object".[58] Parpola collated all the 'writings' unearthed so far, classified them and produced a compendium. The symbolic ideograms appear to convey messages in short texts very often in association with pictures of real and mythological animals and legendary heroes or deities in human form. In addition to the pictographic signs, Parpola[59] writes, "The seals and amulets often contain iconographic motifs, mostly realistic pictures of animals, cultic scenes and human forms". Years of study have revealed about 350 distinct signs in the Indus script. Further, we have been able to make out that the script is generally written from right to left. Obvious symbols such as the vertical strokes ranging in number from 1 to 12 may be taken to indicate numerical values. Despite these little successes, deciphering the Indus script has been in vain in the absence of bilingual texts. Deciphering the Indus script is the most vexing problem of South Asian proto-history.[60]

Parpola[61] speculates on the astronomical significance of some of the Indus seals and links it to the Dravidic language. Positing that the language of the Indus Civilization is Old-Dravidian, he interprets the fish hieroglyph, along with a number of earlier scholars such as Mahadevan,[62] to read as 'men'. '*Min*' in Old Tamil means both 'fish' and 'star'. Both words are derivatives from the Proto-Dravidian root that mean 'to glitter or to sparkle'. The Old Tamil *Caṅkam* text, *Paripāṭal*, compares an earthly river with its fish (*mīn*) to the heavenly Ganges (ecliptic) with its stars (*mīn*), showing that the stars were conceived to be fish (*mīn*) swimming in the waters of the night sky.

The Indus pictogram that contains six short strokes, divided into two lines with an embossed image of fish on its left side, now could be read as 'six stars' which corresponds to the compound 'six-star' (*aṟu-mīṉ*) occurring in Old Tamil texts and denoting the asterism of the Pleiades. This asterism's conjunction with the sun, around 2300 BC, at the vernal equinox was to signify the beginning of the new year 2300 BC. Another Indus seal bears seven strokes and a fish which could then be read as Old Tamil *eḻu-mīṉ*, i.e. 'seven-star' and correspondence with circumpolar 'Great Bear' could be speculated. *Vaṭa-mīn* in Tamil implies, 'North star', that is Polaris. A homonym of *vaṭa* 'north' is *vaṭam* which is 'banyan fig'. *Vaṭam* in Old Tamil also means 'rope', and the tree with a rope-like aerial root system got its name from this association. The 'three-branched fig tree' motif, often found in the painted Indus pottery usually before an image of fish, could then imply 'North Star' in the Dravidian homonymy. Parpola also argues that the ancient cosmogony idea that all the stars are tied by invisible 'ropes' as they rotate around the Pole star is revealed by these images.

Similarly, forceful interpretations are placed to show that Dravidian names of stars and planets, preserved in Old Tamil, occur in Indus inscriptions. Parpola's enthusiasm is not shared by all alike. Possehl[63] comes to the conclusion that none of the proposed models for interpretation of the Indus script stands close scrutiny and satisfaction, and Farmer et al.[64] say that the Indus pictogram is not a script and add further that the "standard view that the Indus Civilization was literate has been an assumption and not a conclusion of previous studies".

While grounds for doubting such interpretations persist, in the present state of the decipherment of the Indus script, Parpola's interpretations are certainly probable. Following Parpola, Ashfaque[65] speculates that the figures of real and mythological animals, bulls, elephants, rams, rhinoceroses, crocodiles, tigers, unicorns, animals with composite heads, etc., found in the Indus seals, "presumed to signify some kind of a crude system of division of the sky by asterisms, roughly marked along the apparent path of the moon as a frame of reference for calendrical reckoning".

Astronomical Import of Orientation

The layout of Indus Civilization towns and cities is taken to imply good astronomical knowledge and possible cosmological beliefs that would have impelled the builders to particularly orient the streets towards a definite direction. The orientation of the main streets in the cities and towns excavated shows a definite preference for an orientation generally towards the cardinal direction. Kenoyer[66] observes, "The long continuity in building orientation and brick size, from Ravi through Kot Diji phase occupations, cannot be coincidental and must be linked to deeply held socio-religious beliefs associated with settlement organization".

The idea of constructing buildings taking into view the cardinal directions survived even as late as the 18th century. The travelogue of Le Gentil, who visited India to observe the transits of Venus in 1761 and 1768, mentions the use of gnomon to fix the cardinal points for laying the foundation, resulting in the perfect alignment of the buildings and monuments to the cardinal points.[67] Finding cardinal directions is not that difficult. A gnomon with a circle of radius equalling the height of the gnomon would give us the east–west line. As the day progressed, the length of the shadow of the gnomon would first decrease, and then after, the local noon commences to increase again. The shadow will touch the circle at two points, one in the morning and another in the evening. If one could place markers at these points and could draw a straight line connecting them, we could obtain an east–west line.

However, the correct orientation of the streets in Mohenjodaro is not perfect east–west but tilted clockwise at about 1°–2° from the north–south line. Wanzke[68] identified this in the layout of the Mohenjodaro's streets, and further excavations at other Indus sites have yielded similar deviations. Wanzke, therefore, proposed that Indus builders established the cardinal directions not from the gnomon shadow but by sighting the heliacal rising of stars at the horizon. He argued that in the case of Mohenjodaro, the builders could have used the profile of the Kirthar Mountains to the city's west as the horizon against which to observe the setting point of the Aldebaran (*Rohin*i) or Procyon, which was north of the cardinal west, exactly matching the orientation of the city's streets. Parpola[69] suggests that Pleiades would have arisen in approximately the same spot in the horizon during each vernal equinox from 2720 to 1760 BC which would have marked the new year in their calendar as well as have provided orientation to the settlement layout.

Jansen,[70] however, cautions that apparent orientation cannot be construed as deliberate and astronomical in nature. He shows that alterations in the orientation of houses and streets can be observed in Mohenjodaro, which shows gradual change as new constructions are made over the earlier layer of the city. His detailed investigation shows that[71] the "orientation of house walls in HR area seems to prove that the north–south orientation shifted toward north-northeast in the course of the period of occupation".

Heliacal Rising of Pleiades

The heliacal rising of a star or of a constellation occurs when it becomes visible above the eastern horizon just before sunrise, after a period conjunction with the sun. The corresponding rising of a celestial body above the eastern horizon at nightfall is its acronychal rising. The Pleiades, being distinctive, were one of the early choices for noting the cyclic motion. While the sun rises in the due east only on two days in a year, Pleiades appear to rise from the east. It is interesting to note that *Śatapatha-Brāhmaṇa* observes that while all other asterisms do move from the eastern quarter, Pleiades remain in the eastern quarter.[72] Baudhāyana's *Śrautasūtra*, dated 700 BCE, states that the Pleiades do not forsake the eastern direction and advises the construction of a scared hut, which requires facing east to use this to find the orientation.[73]

The Pleiades, known as the Seven Sisters in Greece, as the six faces of *Muruga* in Tamilagam or *Kṛttikā* in Vedic Civilization, are an open cluster in the constellation of Taurus. As it is one of the nearest star clusters, visually it is tantalizing and would have been obvious to the naked eye. Appearing brilliant in the night sky during winter in the Northern Hemisphere since

> prehistoric and proto-historic time and all over the world most notably the Pleiades (M 45) and to a less degree the Hyades (Mel 25) had

> attracted great attention among the people. In particular, the Pleiades were important in archaic cultures.[74]

Several cultures all around the world, including the Maori and Australian Aborigines, Chinese, Maya and Aztec and the Native people of North America, have myths associated with it.[75]

Although to a careful observer, more than six stars may be visible on a clear moonless night and six of the bright stars would have been readily visible in the past, devoid of light pollution. In Greek mythology, they are Seven Sisters, daughters of Atlas and Pleione, who were pursued by Orion and rescued by Zeus to be immortalized in the sky. In one of the Hindu myths, originally, the Pleiades were the wives of the Seven Sages (*saptarṣi*, i.e. the Big Dipper) but are now precluded from intercourse with their husbands, as the Seven Sages rise in the north but the Pleiades in the east. Arundhatī (Alcor) was exempted from being separated from her husband Vasistha (Mizar) and can be seen close to that star in the night sky.

A tiny amulet seal at Rehman Dherai at Kot Dijian level has a pair of scorpions and a pair of antelope on the reverse side. In the light of the ancient myths, Ashfaque[76] makes a case to interpret this to represent the era in which the heliacal rising of Orion has fallen out of step with the beginning of a new year around the vernal equinox. Indeed, around 2240 BC during the Mature Harappan period, on the day of spring, the equinox coincided with the heliacal rising of the Pleiades, which perhaps marked the beginning of a new year in the Indus culture. Perhaps earlier, say around 3054 BC, Aldebaran, which rose heliacally at the spring equinox, used to mark the commencement of the New Year. Ashfaque[77] suggests that the myth of the incestuous seduction of Rohini (Aldebaran) by her father Prajapati (Orion) reflects the observed rapprochement between these asterisms over time, and the rejection of Aldebaran in favour of the Pleiades as the marker of the vernal equinox.

Astronomy in Indus Artefacts

Scholars have also looked at the artefacts excavated from Indus sites to tease astronomical implications. Of various artefacts located at Mohenjodaro, a curious doughnut-shaped large stone, which has outer diameters of about 45 cm and thickness of the order of 26 cm, is of particular interest. These perforated stone rings usually made of limestone or calcareous sandstone are the largest and heaviest artefacts recovered at Indus Civilization cities located on the plains. Maula[78] reads them to be instruments designed to keep track of the movement of the sun and identify solstices and the equinoxes. Alternatively, scholars have interpreted it to be Shivite '*yoni*' stones or ceremonial stones associated with cultic tree-worship, or as just plain decorative bases for wooden columns. The sightline connecting specific perforations around the edges of the stone rings are supposed to point to specific directions towards the horizon. When the sun (or any other heavenly object) appears through the sightline, a specific day in the progression of the year could be computed.

A Tamil Sangam literature has a vivid description of an instrument, '*iru kol kuri*', with a stone base and two small poles vertically placed over it. *Pathuppāttu-Nedunalvādai* (72–78)[79] describes the day on which the building of the palace commenced,

> At mid-day when the bright, ray-spreading sun climbed on toward the west side and was high in the vast sky, placed two (vertical) sticks (at noon time) when they did not see the sun's shadow fall on land, learned men, who had read books on construction, noted the cardinal directions, decided where to construct, and on the ground prayed to the Gods and built the palace with many rooms suitable for the well renowned king (lit trns., by author).

It is to be noted that in the poetry word, "*iru kol kuri*", is used as a noun, suggesting an instrument (gadget). When two vertical poles are placed on a plane along a north–south line, then exactly at local noon, the shadows of both the poles would merge, thus readily marking the local noon time. In addition, if the sun is exactly at its zenith, then the poles will not cast any shadows. A palace was to be constructed at Madurai; an ancient South Indian urban centre located about 10° north. The two poles, the poetry says, did not cast a shadow, that is zero shadow days. At Madurai, the sun would reach the zenith on ~April 28 and ~August 16, and we can infer that the construction began on either of these two days.

Excavation of Harappa yielded several circular and wide brick-working platforms with a wooden object embedded in the socket at the centre and projecting about the platform. The platform consists of five concentric rings of burnt bricks set on an edge and one brick in thickness. Wheeler[80] interpreted it to be "a wooden mortar where grain was pounded by one or more workers with long pestles". N. K. Rao[81] observes that the "layout of the bricks in rings is very similar to some of the sundials seen in later years (e.g. *kiranada kalu* of 11th-century sculpture or the Greek sundials)". He suggests that "the central wooden peg (or shaft) casting a shadow to measure the time is one ideal use of such a brick platform"[82] and that edges of bricks and circular layout served as graduation marks on the ground to measure the position of the shadow of the gnomon. The platforms are wide as clear space is needed to see and measure the shadow.

S. R. Rao[83] conjectures that hollow rings with slits on the upper and lower margins are sextant to measure the angle. The shell ring from Dholavira has six slits in the upper margin and six more in the lower margin. The ring from Lothal has four slits in each margin. When lines passing through the opposite slit on both margins are drawn on a plain surface, they cut an angle of 30° in Dholavira and 45° in the Lothal ring. Suspended by a thread or wire, passing through the central hole, this instrument could be used as a sextant.

Out of 1500-odd Indus Civilization sites known today, no structure with definite astronomical implication has been identified so far. A recent computer-simulated study of a structure in Bailey at Dholavira strongly suggests celestial orientations inbuilt into their design. One of the two circular structures faces exact north and this two-room structure may have been constructed specifically in response to the solar geometry at this site, and it is possible that the two circular rooms in the structure were designed for solar observations. If this supposition is correct, then this is the first identified example of a Harappan building that was used specifically for observational astronomy.[84] By assuming a strategic hole in the ceiling and reasonable superstructure, simulations revealed interesting patterns of sunlight on astronomically significant dates.

Megaliths – Potential Sites of Observational Astronomy

Megaliths structures built with large stones are present all over the world. Having mastered working and shaping stones, the Neolithic cultures constructed huge structures, called 'megaliths', incorporating massive stone blocks. The most famous Megalithic site is Stonehenge in the United Kingdom, whose enigmatic stone arrangements have baffled archaeologists for centuries. Historians usually classify megaliths as any large stone object deliberately placed at a specific location for some presumed purpose.

Megaliths in India

In India, megaliths have existed from the Mesolithic period and through the Neolithic period. Mohanty et al.[85] contend that Indian megaliths belong to Iron Age (1500 BC–AD 200), as often iron tools and implements (arrowheads, daggers and domestic vessels) are associated with megaliths settlement. The advent of iron made it possible for humans to prepare strong tools, such as iron plough and further enabled them to chisel shapes into hard rock. Iron technology emerged in South Asia from regional Chalcolithic and Bronze Age traditions by 1000 BC, thereby replacing the dominance of these earlier metallurgical industries. Gandharan Grave Culture of northern Pakistan (1000–300 BC), Pirak and other sites on the Kachi Plain of Pakistan (~900 BC) and Painted Grey Ware culture extending from Baluchistan to Uttar Pradesh in northern India (900–300 BC) clearly provide evidence for iron and iron implements in these archaeological assemblages.

Babington first reported on the megaliths monument, Bangla Motta Paramba, in northern Kerala in 1823. The 'Megalithic complex', dating back to 2000 BC in Kashmir to as recent as AD 50 in South India, associated with Iron Age culture has been discovered through field survey.

Summarizing the evidence from the research carried out so far, the megaliths of peninsular India can be tentatively dated from the end of 2000 BC to the middle of AD 1000.

Within South India, northern Karnataka and Andhra Pradesh show an early beginning, while those of Tamil Nadu and Kerala are in the later stages. The Megalithic tradition continued to exist even during the 4th and 5th centuries AD in Tamil Nadu and Kerala. Srinivasan[86] compared the South Indian megaliths with that of references to burial customs in early Tamil Sangam texts and contended that South Indian Megalithic culture was well established before 300 BC. Vindhyan region megaliths are a mixture of pre-Iron Age megaliths and proper Iron Age megaliths. The pre-Iron Age megaliths are dated from 1500 BC to 1000 BC, while the Iron Age megaliths are between 800 BC and 300 BC. Megaliths, at Gufkral Kashmir, are around 1500 BC, while Gagrigol, in the Kumaon region, is around 2500 BC, indicating slightly earlier Megalithic culture in Northern India. S. B. Deo[87] catalogues 1400 Megalithic sites in the subcontinent; however, 1116 of them are in peninsular India. A recent survey by Mohanty and Selvakumar[88] shows about 2200 Megalithic sites in South India alone, and not many of them have been excavated extensively. Since 1823, about 239 have been excavated in some way, often by treasure hunters rather than by trained archaeologists. Only about 30 of them have been undertaken by trained archaeologists, and the human remains have been examined using the scientific method.

Megalith Culture

The Megalithic period witnessed a population increase and emergence of craft specialization and an increased interaction between various regions in South India. A distinct Chalcolithic period in northern India is discernible; however, in South India, the Mesolithic gave way to a more mixed economy based on agro-pastoral production using iron implements. It was a transitory phase from which settled life and urbanization emerged. Agriculture and cattle pastoralism are seen together along with the use of a wide variety of fabrics and ceramics. Hunting gathering also seems to be widely prevalent from the archaeological tool remains and paintings of that period.

Most of the South Indian megaliths were located in regions which could command water, minerals and arable land. Moorti[89] identifies 26 large settlements, each having a population of the order of 1000 that could have served as regional centres of trade. Long-distance trade gradually increased in the late phase of the Megalithic cultures, leading to the rapid distribution of artefacts and ideas, as has been evidenced by the beads and ornaments of the period. Evidence from the early historic Megalithic sites, like Kodumanal, indicates the active participation of Megalithic communities in the trade even with faraway places such as Rome.

Astronomical Significance of Megaliths

Baity et al.[90] convincingly established that at least some of the megaliths in Europe have astronomical significance. The stone arrangements are supposed to provide clear sightlines for the azimuth of the rising or setting sun (or moon, or an important bright star) during specific astronomical occasions such as equinoxes.[91] For an observer, standing at a specifically defined location, alignment of rocks or the profile of the rock in the background of the horizon constitute sightlines. World over archaeo-astronomers are finding that the Megalithic stone alignments have a probable astronomical purpose.[92] The Megalithic period, in Indian culture, is marked by the presence of huge stones arranged in a definitive pattern. It is possible that many or all of such megaliths were erected for ritualistic purposes, but a fair case has been made by scholars that the primary role of many of them was to keep track of the movements of astronomical bodies.

The megaliths usually are spectacular, elaborate, spread over a larger area of several hundred square metres and seem to have been well planned before being erected. The early megaliths were mostly large stones put at specific locations, but the megaliths of the mature phase tend to be dressed in boulders and chiselled stones. Moorti[93] classifies most of these Megalithic monuments into two broad categories: Sepulchral and Non-Sepulchral. Sepulchral monuments are those which contain (or contained) the physical remains of dead human beings, and non-sepulchral are those which were ceremonial or memorial in nature. The menhir and other types of later period show work before the stone was used for the specific purpose of non-sepulchral markers.

The sepulchral group of megaliths is clearly intended as the final resting place of the deceased. Its purpose may be more than just a burial marker for the dead and could have socio-ritual functions and be linked to cosmologeny of the culture. Nevertheless, the primary function is to host one or more dead bodies. On the other hand, the purposes for which the non-sepulchral monuments were erected are not clear. It may have been ritualistic, commemorative or may have had other purposes which we can only speculate. It may be possible to imagine dolmen serving as an ancestral shrine, but the purpose of the arduous effort to erect a stone-avenue or alignment is not readily apparent.

South India

Examining the sepulchral megaliths in South India, Sundara[94] observed that the direction of the non-circular burials is often in the cardinal directions. In addition, many of these sepulchral megaliths have ancillary surface markers pointing to (usually) the northern direction. In a large number of cases, the bodies aligned either north–south (with the head in the north) or east–west (with the head in the east).[95]

After surveying what at the first glance appeared as haphazard clusters of megalith menhirs in the south Canara region of Karnataka state in South India, Srikumar Menon et al.[96] show five monuments in the region that exhibit "a positive correlation of sight-lines with sunrise and sunset points on the horizon for both summer and winter solstices" demonstrating astronomical implications. In like manner, the stone alignment at Byse also indicates astronomical significance.[97] K. P. Rao reports the astronomical significance of many South Indian megaliths.[98] For example, an ancient sky map assignable to the Megalithic period in Mudumal in Mahbubnagar district, Andhra Pradesh is reported by him in detail.[99] This Megalithic complex has menhirs, cists and stone circles. The menhirs "are planted in such a way that the rows are exactly in line with the rising and setting sun on the days of solar significance", and that on "20-21 June (summer solstice) and 21-22 December (winter solstice) revealed that specific alignments are aligned to the sun on morning and evening". Further, the rows are "planned in such a way that it gets aligned to the sun at specific time and date. If one row gets aligned in the morning, another row gets aligned in the evening" and infer that the "complex is planned in such a manner that it is possible to determine the commencement of seasons or workout a sort of calendar on the basis of sun's position in relation to these monuments".[100] K. P. Rao[101] interprets that the pattern of the cup marks in the stone circle resembles the star pattern of the Ursa Major constellation, in particular Big Dipper. However, Rao et al.[102] contest the claim and argue that the "rock surface contains about fifty cup marks, not just seven" and assert that "any resemblance to the Great Bear constellation is imaginary. There is no indication that these marks are even depictions of stars".

Examining the stone alignments of the Brahmagiri Megalithic stone circles (c. 900 BC), N. K. Rao[103] contends that they exhibit clear astronomical orientation. He argues that the site from the centre of a circle to an outer tangent of another circle and points out the directions of the sunrise and moon rise at the time of solar and lunar solstices and equinox. Furthermore, the study of the sepulchral stone circles at Junapani also indicates alignment to cardinal directions.[104] Thus, even in the case of sepulchral megaliths, there appears to be some connection between the sepulchral structures and astronomy possibly through cosmogony.

Central Region Megaliths

The megaliths in the Vidarbha region of Central India have been catalogued by Suvrathan.[105] Of the 89 Megalithic sites in this region, 51 are concentrated around the Nagpur region.[106] Further, predominantly, the megaliths of this region are of stone circle types. While 54 sites have only stone circles, 4 sites have a dolmen, while 1 has a dolmen and a stone circle. Khairwada site, with a 1400 stone circle, is the largest. Jamkhedkar found the habitation deposits on this site that spans from the Megalithic to the early historical period. The second largest site is Junapani which has 150 stone circles. In this site, excavated by Thapar,[107] all these circles appear to belong to the same period.[108] One of the unusual features of these stone circles was that some of the stones in the circle had cup-like marks of the typical size of a few centimetres definitely chiselled by a human hand. Vahia et al.,[109] in their field survey that mapped 56 stone circles, found that the location of these stones with cup marks was not random but seemed to have a definite pattern. The specific locations of the cup marks along the stone circles and their orientation suggest that "they were probably designed to mark out specific locations in the sky that corresponded to rising and setting time of specific stars associated with important changes in seasons and especially with the arrival of monsoon". In a given circle, the "cup-marks tend to be in either a straight line or in the form of the '+' sign. In addition, the line or one of the lines of the '+' sign tend to be radial or tangential to the circle".[110]

Orientation of Menhirs

The utility of most of the non-sepulchral monuments has often been speculated upon and the purpose is often difficult to define. They do not have the dead buried under them and hence possibly are not associated with memorial stones for the dead. A number of scholars have suggested that such stone structures requiring substantial human labour are astronomical in nature. These stone arrangements could serve as calendar devices for instance. The definite pattern in the placement of the stones or cup marks on the stones, the orientation of menhirs; in short, the arrangements are "far too complex to admit of simple explanations. Almost all the megaliths have alignments which are either north-south or east-west with one prominent marker stone or a porthole in one stone pointing to either north or east".[111]

Investigating the Buddhist stupas of Sanchi, erected around the 2nd century BC by the Sunga kings, N. K. Rao[112] argues that it is oriented towards moonrise and sunset on the day of Buddha Purnima and further that the stupa 1 and stupa 2 could have served as a luni-solar astronomical calendar. The cardinal arrangements of places of worship in Hindu, Buddhist and other Indian religious cultures are well known. Could it be that these non-sepulchral monuments have markers pointing to specific directions or markers indicating sightlines for rise-set points of the sun, moon and perhaps certain specific constellations?

Stone Avenues

One of the spectacular non-sepulchral monuments is what is described as stone alignments or stone avenues. A large number of stones, often around thousands, are arranged into a grid pattern rather than circles. Mostly these monuments have no human remains buried under them. Stone alignments at Vibhutihalli near Shahpur Karnataka consist of boulders rolled down from the nearby granite hill and arranged into a specific pattern. About 1000 stones are arranged in a diagonal grid; however, today, some of these stones are missing. The more grandeur stone alignments are at Hanamsagar, northern Karnataka, probably made out of 2500 stones, arranged in diagonal grid pattern. Archaeologists earlier speculated that these arrangements were markers for royal camping ground or place for markets or fairs. Some even proposed that these were prophylactic stones for preventing sickness in cattle.

Allchin[113] has catalogued many stone alignments in northern Karnataka and Andhra Pradesh. He found most of these to be grids or staggered (diagonal) grids and "Hyderabad alignments are rows or diagonal rows of rough-hewn stones or boulders set out with mathematical precision and oriented to the cardinal points".[114] They were found to be either aligned to the cardinal directions or around 15° off from the cardinal directions. Rao and Thakur[115] argue that the alignment at Vibhuthihalli is astronomical in nature, i.e. aligned to points of solar and lunar significance on the local horizon. Non-sepulchral Megalithic monument at Nilurallu in Murardoddi is another avenue-type megalith with stones 12–16 feet high arranged into a pattern that appears squarish. N. K. Rao et al.[116] show that these rows of stones are "distinctly aligned to calendrically important events, like to sunrises and sunsets during both the equinoxes and the solstices".

Likewise, investigating the stone alignments of Hanamsagar, Paddayya[117] and Rao[118] suggest astronomical imperatives for the layout of the stones. Although another study of Hanamsagar[119] shows that "the stone rows may be curving away like concentric arcs" and suggest that "shape attributed to the alignment is questionable and a proper aerial survey or high-resolution satellite image" are required to obtain better evidence of sightlines. Menon et al.[120] contend that stone alignments/avenue-type megaliths could have had a possible use as calendar devices. They surveyed the clusters of menhirs in the south Canara region of Karnataka state in South India and have identified five monuments demonstrating astronomical implications having a "positive correlation of sight-lines with sunrise and sunset points on the horizon for both summer and winter solstices".[121]

Menon[122] surveyed 33 sites (20 in Karnataka, 8 in Kerala, 3 in the Vidarbha region of Maharashtra and 1 in Tamil Nadu), carried out orientation surveys in 12 and partial surveys at 4 sites and extensive surveys in another 4 sites. Some showed south-facing and a few others east–west facing orientations, while there were some which showed no preferred orientation at all, revealing regional variation in the orientation. He concludes that while the earliest Megalithic types had no preferred orientations to cardinal directions or were exclusively north–south oriented, the chronologically later types and architecturally more developed monuments preferred east–west orientations.[123] Their investigation of the non-sepulchral megalith type called menhirs/alignments (or avenues) provides interesting clues, and the unique Megalithic avenue-type monument in southern coastal Karnataka consisting of several menhirs that are erected on end in specific patterns are aligned to the extreme rising/setting points of the sun and moon on the local horizon.[124] Five sites, i.e. Nilaskal, Byse, Heragal, Mumbaru and Aaraga Gate, show astronomical sightlines and probably had astronomical implications.

Megaliths at the Nilaskal Byana (field with the standing stones), a complex of more than 100 menhirs and their remnants spread over around 30,000 square metres, are an enigma. Menon[125] shows that the menhirs are found only on the "east facing slope and none on the other side of the ridge". Further, several stones have notches cut into them, suggesting a peep-hole to permit the viewing of other menhirs that constitute the sightlines against the horizon. Menon et al.[126] show intentional alignments to the solar points and possibly lunar points of the Byse menhir complex. Apart from the solstice sightlines found in many other Megalithic sites, Byse seems to have specific markers for the setting points of important stars. Rather than a central observation location, as in the case of many well-known European megaliths, this site has "multiple sightlines with shadows of several stones falling on other marking stones during days of astronomical importance".[127]

Problems in Inference

To ferret out the astronomical knowledge and cosmogony from the form and orientation of megaliths is a daunting task. It is easy to mistake chance alignment and conjure deliberate orientation and imagine sightlines. Ruggles[128] perceptively notes that the "central question [is] ... whether each given set of putative sightlines could quite adequately be explained away as chance occurrence". Heggie[129] points out that

> the great bulk of the evidence on Megalithic astronomy consists of orientation, alignments of 'lines' directed towards the place on the horizon where ... [a] variety of astronomically significant positions [is seen] ... one must consider the possibility that it is only by chance one finds some sites to be astronomically oriented ... [and] we can be virtually certain that coincidences occur ... it seems wise to dismiss any apparently astronomical orientations as coincidences unless there is evidence to the contrary.

Further, as most of these monuments are unprotected, they are subject to heavy disturbance by development activities. Shifting of the stones in megaliths by subsequent generations to clear an area may provide us with erroneous perceptions.

Archaeo-astronomy reveals potentially how these stone alignments could have been used for astronomical observation during the Megalithic times. However, we have very little direct evidence of its use for astronomical purposes. The arbitrariness of some of these directions selected (and not others) for the observation ascribed to particular sites by current-day archaeo-astronomers is a principal ground of criticism. When a large number of menhirs or boulders are present, permutation and combination between them will give umpteen sightlines. In addition, how far can we be certain of the current position of a monument to its placement at the time of its erection and use? Earth tremors, agricultural activity or just vandalism could have moved the stone from its original location. Archaeological surveys indicate many Megalithic stones have been moved in the past and re-erected. If indeed the stones are realigned from the first placement and still we are reading the sightline layout in the arrangement, then it is we who are reading a purpose which was not intended. Even assuming that the stone placements are undisturbed for many centuries, thirdly, thick vegetation growth in the east or west of the observatory could effectively hide the horizon and mar the effective observation. Today, open farmlands are much more common due to the clearing of forests for agriculture. However, during the time of megaliths, although agriculture was probably practised, as much of the forest clearance is in the historical period, we are clueless about the extent of local vegetation around the sites.[130]

Observatory of the Ancient World

So far, some or all of the megaliths have astronomical or cosmogony implications and they are the first creative scientific pursuit. Indeed, the discovery of fire and the fabrication of the wheel were revolutionary; they turned the course of history. However, such inventions did not provide an understanding of the underlying nature but were empirical adaptations. Although astronomy worked in the Megalithic observatories may have had direct use in predicting the onset of monsoon, completion of the year and so on. Astronomy went beyond these direct utilitarian and, over time, went on to work out longer term periodicity, such as the eclipse cycle and so on. In this sense, megaliths were an inceptive 'astronomical observatory' of the ancient world.

However, one hastens to caution the reader to the notion of 'observatory' which is not the same as the modern meaning, a structure erected primarily for the purpose of astronomy. Belmonte[131] notes that a

> certain monument, or a spot clearly signalled by artificial elements, could be considered an observatory if it can be demonstrated that the astronomical components on site had 'a practical purpose for the living' that does not need to be a purely scientific one. For example, it could have been used to regulate time for the creation and control of a calendar. Sites for the dead, such as Newgrange or the pyramids, or for the gods like astronomically aligned temples such as Karnak or the Pantheon, should not be considered in this category.

In so far as the Megalithic stone arrangements are actually 'astronomical observatories' of the ancient world, they often indicate not just awareness of the cardinal directions, but also a sophisticated knowledge of the cultures to make the connection between seasons and shifts in the sunrise points. From the layout of sightlines evident in the Megalithics, we can infer that this culture was aware of the significant events, such as solstices and equinoxes in the cycle of the sun and major and minor standstills of the moon (called Lunistices) in the moon cycle.

Megalithics represent the first example of applied science since the architecture of stone arrangements is astronomy. They are also the first major scientific tools in the sense that they are not empirical constructions that are just utilitarian. Inducting from observations carried over many years and correlating visible markers like sunrise points with complex cycles, a set of ideas are formalized. Formalized understandings of the movement of heavenly objects in the background of the horizon are used to decide the stone arrangements. Time periods beyond days and months could be measured by these archaic astronomical observatories. In serving as a calendrical device, megaliths provide temporal order beyond the immediate and day-to-day existence, elevating human culture much beyond that of animal existence.

Notes

1 Bidare Venkatasubbiah Subbarayappa and K. Venkateswara Sarma, *Indian Astronomy: A Source-Book, Based Primarily on Sanskrit Texts*, Nehru Centre, Bombay, 1985, p. 1.
2 *Ibid.*, p. xx.
3 A. K. Bag, "Early System of Nakṣatras, Calendar, Antiquity of Vedic and Harappan Traditions", *Indian Journal of History of Science*, Vol. 50, No. 1, 2015, pp. 1–25; Yukio Ôhashi, "Astronomy of the Vedic Age", in Clive L. N. Ruggles (ed.), *Handbook of Archaeoastronomy and Ethnoastronomy*, Springer, New York, 2015, pp. 1949–1958 (p. 1951).
4 Yukio Ôhashi, *ibid.*, p. 1952.
5 T. S. Kuppanna Sastry and K. V. Sarma (eds.), *Vedanga Jyotisa of Lagadha* in its Rk and Yajus Recensions: With the translation and notes of Prof. T. S. Kuppanna Sastry, Indian National Science Academy, New Delhi, 1985, p. 11.
6 Yukio Ôhashi, "Development of Astronomical Observation in Vedic and Post-Vedic India", *Indian Journal of History of Science*, Vol. 28, No. 3, 1993, pp. 185–251; Yukio Ôhashi, "On Vedāṅga Astronomy: The Earliest Systematic Indian Astronomy", in T. Nakamura, W. Orchiston, M. Sôma and R. Strom (eds.), *Mapping the Oriental Sky, Proceedings of the Seventh International Conference on Oriental Astronomy*, National Astronomical Observatory of Japan, Tokyo, pp. 164–170.
7 Manu V. Devadevan, "Śulbasūtras", in *Encyclopaedia of the History of Science, Technology and Medicine in Non-Western Cultures*, Springer, Netherlands (Dordrecht), 2016, pp. 4079–4080.
8 George Abraham, "Gnomon in India", in *Encyclopaedia of the History of Science, Technology and Medicine in Non-Western Cultures*, Springer Netherlands, Dordrecht, 2016, pp. 2126–2129.
9 Yukio Ôhashi, "Development of Astronomical Observation in Vedic and Post-Vedic India", *op. cit.*, pp. 185–251.
10 Kenneth A. R. Kennedy, "Middle and Late Pleistocene Hominids of South Asia", in *Humanity from African Naissance to Coming Millennia: Colloquia*

in Human Biology and Palaeoanthropology, (Monografie.Scienze; 1), Firenze: Johannesburg (ZA): Firenze University Press, Witwatersrand University Press, Johannesburg, 2001, pp. 167–174; see also Bridget Allchin and Raymond Allchin, *The Rise of Civilization in India and Pakistan*, Cambridge World Archaeology, Cambridge University Press, Cambridge, 1982; Burjor Avari, *India: The Ancient Past: A History of the Indian Subcontinent from c. 7000 BCE to CE 1200*, Second edition, Routledge, New York, 2016; Dorian Q. Fuller, "Non-Human Genetics, Agricultural Origins and Historical Linguistics in South Asia", in Michael D. Petraglia and Bridget Allchin (eds.), *The Evolution and History of Human Populations in South Asia: Inter-Disciplinary Studies in Archaeology, Biological Anthropology, Linguistics and Genetics*, Springer, Netherlands, 2007; Kenneth A. R. Kennedy, *God-Apes and Fossil Men,* University of Michigan Press, Ann Arbor, MI, 2000; Hasmukhlal Dhirajlal Sankalia, *The Prehistory and Protohistory of India and Pakistan,* University of Bombay, Bombay, 1962.

11 Nicole Boivin, "Orientalism, Ideology and Identity", *Journal of Social Archaeology*, Vol. 5, No. 2, Sage Journals, 2005, pp. 225–252; Shereen Ratnagar, "Archaeology at the Heart of a Political Confrontation", *Current Anthropology*, Vol. 45, No. 2, The University of Chicago Press, USA, 2004, pp. 239–259.

12 Giulio Magli, *Archaeoastronomy*, Undergraduate Lecture Notes in Physics, Springer International Publishing, Cham, Switzerland, 2016, p. 3.

13 Edwin Charles Krupp, "Archaeoastronomy", in John Lankford (ed.), *History of Astronomy: An Encyclopaedia*, Routledge, New York, 2011, pp. 21–30 (p. 21).

14 Rajan Gurukkal and Raghava M. R. Varier, *Cultural History of Kerala,* Dept. Of Cultural Publications, Govt. of Kerala, Thiruvananthapuram, Kerala, 1999, p. 145.

15 Anthony F. Aveni, "Archaeoastronomy", *Advances in Archaeological Method and Theory*, Vol. 4, Springer, 1981, pp. 1–77 (p. 1).

16 *Ibid.*, p. 2.

17 Clive L. N. Ruggles (ed.), *op. cit.,* pp. 15–30.

18 Hugh Thurston, *Early Astronomy*, Springer, New York, 1996, pp. 178–189.

19 James Evans, *The History & Practice of Ancient Astronomy,* Oxford University Press, New York, 1998, pp. 3–72.

20 Anthony F. Aveni, *Empires of Time: Calendars, Clocks, and Cultures*, Basic Books Inc., New York, 1989, pp. 38–81.

21 See for an exhaustive and recent study of astronomical practices and beliefs among many tribes in Central India, Ganesh Halkare, Mayank N. Vahia and Wayne Orchiston, "Astronomy of Some Indian Tribes", in Wayne Orchiston, Aniket Sule, and Mayank N. Vahia (eds.), *The Growth and Development of Astronomy and Astrophysics in India and the Asia-Pacific Region*, *Astrophysics and Space Science Proceedings*, Vol. 54, Springer, Singapore, 2019, pp. 423–31; Mayank N. Vahia and others, "Astronomy of Two Indian Tribes: Banjaras and Kolams", *Journal of Astronomical History and Heritage*, Thailand, Vol. 17, No. 1, 2014, p. 20.

22 Martin P. Nilsson, *Primitive Time-Reckoning: A Study in the Origins and First Development of the Art of Counting Time among the Primitive and Early Culture Peoples*, Oxford University Press, Lund & Oxford, 1920, p. 2.

23 Clive L. N. Ruggles, "Calendars and Astronomy", in Clive L. N. Ruggles (ed.), *op. cit.,* Springer, New York, 2015, pp. 15–30 (p. 15).

24 Umed Bulbulshoev, Karim-Aly Kassam and Morgan Ruelle, "Ecology of Time: Calendar of the Human Body in the Pamir Mountains", *Journal of Persianate Studies*, Brill Publishers, Vol. 4, No. 2, 2011, pp. 146–170.

25 Anthony F. Aveni, *Empires of Time: Calendars, Clocks, and Cultures*, *op. cit.*, pp. 167–304.

26 Stanisław Iwaniszewski, "Concepts of Space, Time, and the Cosmos", in Clive L. N. Ruggles (ed.), *op. cit.*, pp. 3–14.

27 N. K. Sundareswaran, "The Misinterpretation of the Aryabhata I's Theory of Rotation of the Earth", *Sri Venkateswara University Oriental Journal*, XLIII.1, 2000, pp. 67–76.

28 Bidare Venkatasubbiah Subbarayappa and K. Venkateswara Sarma, *op. cit.*, pp. 11–12.

29 David H. Kelley and Eugene F. Milone, *Exploring Ancient Skies: A Survey of Ancient and Cultural Astronomy*, 2nd edition, Springer, New York, 2011, p. 3.

30 Bradley E. Schaefer, "Heliacal Rise Phenomena", *Journal for the History of Astronomy*, Sage Journal, UK, Vol. 18, No. 11, 1987, pp. 19–33.

31 Trevor M. Leaman, Duane Willis Hamacher and Mark Carter, "Aboriginal Astronomical Traditions from Ooldea, South Australia, Part 2: Animals in the Ooldean Sky", *Journal of Astronomical History and Heritage*, Vol. 19, No. 1, Thailand, 2016, pp. 61–78.

32 Asko Parpola, "History of Science in South Asia", *History of Science in South Asia (HSSA Journal)*, Vol. 1, Canada, 2013, pp. 21–78 (p. 21).

33 W. Randolph Kloetzli, "Maps of Time-Mythologies of Descent: Scientific Instruments and the Purāṇic Cosmograph", *History of Religions*, Vol. 25, No. 2, The University of Chicago Press, 1985, pp. 116–147; R. N. Iyengar, "Dhruva the Ancient Indian Pole Star: Fixity, Rotation and Movement", *Indian Journal of History of Science*, Vol. 46, No. 1, INSA, New Delhi, 2011, pp. 23–39 (pp. 26–27).

34 David Pingree, "The Purāṇas and JyotiḥŚāstra: Astronomy", *Journal of the American Oriental Society*, Vol. 110, No. 2, American Oriental Society, USA, 1990, pp. 274–280.

35 Syed Mohammad Ashfaque, "Astronomy in the Indus Valley Civilization: A Survey of the Problems and Possibilities of the Ancient Indian Astronomy and Cosmology in the Light of Indus Script Decipherment by the Finnish Scholars", *Centaurus*, Vol. 21, No. 2, United Kingdom, 1977, pp. 149–193 (p. 157).

36 B. G. Sidharth, "The Calendric Astronomy of the Vedas", *Bulletin of the Astronomical Society of India*, Hyderabad, Vol. 26, 1998, pp. 107–112.

37 Alexander Marshack, "Upper Palaeolithic Notation and Symbol: Sequential Microscopic Analyses of Magdalenian Engravings Document Possible Cognitive Origins of Writing", *Science,* Vol. 178, No. 4063, USA, 1972, pp. 817–828; Alexander Marshack, "Cognitive Aspects of Upper Palaeolithic Engraving", *Current Anthropology*, Vol. 13, Nos. 3/4, The University of Chicago Press, USA, 1972, pp. 445–477.

38 Jan Gyllenbok, *Encyclopaedia of Historical Metrology, Weights, and Measures*, Volume 1, Science Networks, Historical Studies 56, Springer International Publishing (Birkhäuser), Cham, Switzerland, 2018, p. 239.

39 Brian Hayden and Suzanne Villeneuve, "Astronomy in the Upper Palaeolithic?", *Cambridge Archaeological Journal*, Vol. 21, No. 3, Cambridge University Press, United Kingdom, 2011, pp. 331–355 (p. 331).

40 Francesco D'Errico, "Palaeolithic Lunar Calendars: A Case of Wishful Thinking?", *Current Anthropology*, Vol. 30, No. 1, The University of Chicago Press, USA, 1989, pp. 117–118.

41 Alexander Marshack and Francesco D'Errico, "On Wishful Thinking and Lunar 'Calendars'", *Current Anthropology*, Vol. 30, No. 4, The University of Chicago Press, USA, 1989, pp. 491–500.

42 Sir Richard C. Temple, *The Andaman and Nicobar Islands: Report on the Census*, Census of India, 1901, Vol. III, Office of the Superintendent of Government Printing, Calcutta, India, 1903, p. 81.
43 Mayank N. Vahia and others, "In Situ Studies of Astronomy of the Nicobarese Tribe of Nicobar Islands", *Heritage: Journal of Multidisciplinary Studies in Archaeology*, Vol. 6, Kerala, 2018, pp. 1013–1038 (p. 1019).
44 N. Kameswara Rao, "Aspects of Prehistoric Astronomy in India", *Bulletin of the Astronomical Society of India*, Hyderabad, Vol. 33, 2005, pp. 499–511.
45 See, for a detailed description of the artefacts, Alexander Marshack, *The Roots of Civilization: The Cognitive Beginnings of Man's First Art, Symbol and Notation*, Moyer Bell, New York, 1991, pp. 140–142.
46 Adam Brumm, Nicole Boivin and Richard Fullagar, "Signs of Life: Engraved Stone Artefacts from Neolithic South India", *Cambridge Archaeological Journal*, Cambridge University Press, United Kingdom, 16.2, 2006, pp. 165–190 (p. 180).
47 Cited in Amalananda Ghosh, *An Encyclopaedia of Indian Archaeology* (2 Vols.), Vol. 1, Brill, Leiden, 1990, pp. 34–35.
48 N. Kameswara Rao, *op. cit.*, pp. 499–511.
49 *Ibid.*
50 Hans Kumar and Lotika Varadarajan, "Ap, The Canoe of Chowra IslandAp, the Canoe of Chowra IslandAp, La Canoa de La Isla Chowra", *Techniques & Culture*, 2001, pp. 85–120.
51 Mayank N. Vahia and Nisha Yadav, "The Origin and Growth of Astronomy, as Viewed from an Indian Context", in Wayne Orchiston, Tsuko Nakamura, and Richard G. Strom (eds.), *Highlighting the History of Astronomy in the Asia-Pacific Region, Proceedings of the ICOA-6 Conference*, Springer-Verlag, New York, 2011, p. 64.
52 *Ibid.*, p. 68.
53 Naseer Iqbal, Mayank N. Vahia, Ajaz Ahmad and others, "The Prehistoric Meteor Shower Recorded on a Palaeolithic Rock", *Journal of Astronomy and Astrophysics*, Special Issue, *NRIAG*, Egypt, 2008, pp. 469–475.
54 Hrishikesh Joglekar and others, "Oldest Sky-Chart with Supernova Record", *Journal of Indian Archaeological Society*, New Delhi, Vol. 4325, 2006, pp. 1–7.
55 Naseer Iqbal, Mayank N. Vahia, Tabasum Masood, and others, "Some Early Astronomical Sites in the Kashmir Region", *Journal of Astronomical History and Heritage*, Thailand, Vol. 12, No. 1, 2009, pp. 61–65.
56 Asko Parpola and P. Koskikallio, *South Asian Archaeology, Proceedings of the Twelfth International Conference of the European Association of South Asian Archaeologists* held in Helsinki University, 5–9 July, 1993 (Association of South Asian Archaeologists in Western Europe), Annales Academiae Scientiarum Fennicae, Series B, Suomalainen Tiedeakatemia, 1994, p. 512.
57 Gregory L. Possehl, *The Indus Civilization: A Contemporary Perspective*, AltaMira Press, Walnut Creek, CA, 2002, p. 64; see also Tim Dyson, "A Population History of India From the First Modern People to the Present Day", *Journal of Chemical Information and Modeling*, Oxford University Press, Oxford, 2018.
58 Asko Parpola, *Deciphering the Indus Script,* Cambridge University Press, New York, 1994, p. 3.
59 *Ibid.*
60 Asko Parpola, "The Indus Script: A Challenging Puzzle", *World Archaeology*, Vol. 17, No. 3, Taylor & Francis Ltd., 1986, pp. 399–419.
61 Asko Parpola, *A Dravidian Solution to the Indus Script Problem,* Central Institute of Classical Tamil, Chennai, 2010; Asko Parpola, *Religion Reflected*

in the Iconic Signs of the Indus Script: Penetrating Into Long-Forgotten Picto+graphic Messages, Brill, 1988, p. 15.

62 Mahadevan Iravatham, "Dravidian Parallels in Proto-Indian Script", *Journal of Tamil Studies*, Vol. 2, No. 1, 1970, p. 179 ff.

63 Gregory L. Possehl, *op. cit.*, pp. 134–135.

64 Steve Farmer, Richard Sproat and Michael Witzel, "The Collapse of the Indus-Script Thesis: The Myth of a Literate Harappan Civilization", *Electronic Journal of Vedic Studies*, Vol. 11, No. 2, 2004, pp. 19–57.

65 Syed Mohammad Ashfaque, *op. cit.*, p. 160; Syed Mohammad Ashfaque, "Primitive Astronomy", *Nature*, UK, Vol. 169, No. 4298, 1952, p. 443.

66 Jonathan Mark Kenoyer, "Measuring the Harappan World: Insights into the Indus Order and Cosmology", in Colin Renfrew and Iain Morley (eds.), *The Archaeology of Measurement: Comprehending Heaven, Earth and Time in Ancient Societies*, Cambridge University Press, Cambridge, 2010, pp. 106–122 (p. 114).

67 Cited in Quintin Craufurd, *Sketches Chiefly Relating to the History, Religion, Learning, and Manners, of the Hindoos with a Concise Account of the Present State of the Native Powers of Hindostan*, T. Cadell, London, 1790, pp. 214–215.

68 Holger Wanzke, "Axis Systems and Orientation at Mohenjo-Daro", in *Report on Field Work Carried out at Mohenjo-Daro Interim Reports*, Vol. 2, Pakistan, 1983–84, IsMEO-Aachen University Mission, Edited by M. Jansen and G. Urban, Istituto Italiano per il Medio ed Estremo Oriente, Roma, 1987, pp. 33–44.

69 Asko Parpola, *Deciphering the Indus Script*, *op. cit.*, p. 204.

70 Michael R. N. Jansen, "Mohenjo-Daro: Type Site of the Earliest Urbanization Process in South Asia", in *Studies in the History of Art, 31, Symposium Papers XV: Urban Form and Meaning in South Asia: The Shaping of Cities from Prehistoric to Pre-colonial Times*, 1993, pp. 35–51.

71 *Ibid.*

72 Sankara Balkrishna Dikshit, "The Age of the Śatapatha-Brāhmaṇa", *The Indian Antiquary*, Bombay, Vol. xxiv, 1895, pp. 245–246.

73 C. G. Kashikar (ed.), *Baudhāyana-Śrauta-Sutra (Critically Edited and Translated)*, 4 Vols, Indira Gandhi National Centre for the Arts, Motilal Banarsidass, 2003, p. 1601.

74 Michael A. Rappenglück, "Hyades and Pleiades as a Kind of "Clock" in the Astronomy of Archaic Cultures – The Function and Representation of these Open Clusters in Prehistoric and Protohistoric Time as Well as among Indigenous People", *Astronomische Nachrichten*, Germany, Vol. 328, No. 7, 2007, p. 692.

75 Lynn Ceci, "Watchers of the Pleiades: Ethnoastronomy among Native Cultivators in Northeastern North America", *Ethnohistory*, USA, Vol. 25, No. 4, 1978, pp. 301–317; see also Munya Andrews, *The Seven Sisters of the Pleiades: Stories from around the World*, Spinifex Press, North Melbourne, VIC, 2004.

76 Syed Mohammad Ashfaque, *Astronomy in the Indus Valley Civilization*, *op. cit.*, p. 160.

77 Syed Mohammad Ashfaque, "Primitive Astronomy in the Indus Civilization", in Jonathan Mark Kenoyer (ed.), *Old Problems and New Perspectives in the Archaeology of South Asia*, Department of Anthropology, University of Wisconsin, Madison, WI, 1989, pp. 207–215.

78 Erkka Maula, "The Calendar Stones of Mohenjodaro", in Interim Reports Vol. 1: *Reports on Field Work Carried Out at Mohenjo-Daro, Pakistan*, 1983–84, IsMEO-Aachen University Mission, Edited by M. Jansen and G. Urban, Istituto Italiano per il Medio ed Estremo Oriente, Roma, 1984, pp. 159–170.

79 U. Ve. Saminathaiyer, *Pattuppāṭṭu Mūlamum Uraiyum* (in Tamil), kamarṣiyala accukkūṭam, Chennai, 1918, p. 324.
80 R. E. M. Wheeler, "Harappa 1946: The Defences and Cemetery R37", *Ancient India*, ASI, New Delhi, 1947, pp. 58–130 (pp. 77–78).
81 N. Kameswara Rao, *op. cit.*, p. 506.
82 *Ibid.*
83 Shikaripura Ranganatha Rao, "A Navigational Instrument of the Harappan Sailors", *Marine Archaeology*, NIO, Goa, 3 July, 1992, pp. 61–66.
84 Mayank N. Vahia and Srikumar M. Menon, "A Possible Harappan Astronomical Observatory at Dholavira", *Journal of Astronomical History and Heritage*, Thailand, Vol. 16, No. 3, 2013, pp. 261–268 (p. 267).
85 R. K. Mohanty and V. Selvakumar, "The Archaeology of the Megalithic in India: (1947–1997)", in S. Settar and Ravi Korisettar (eds.), *Indian Archaeology in Retrospect: Prehistory - The Early Archaeology of South Asia*, Vol. 1, ICHR & Manohar, New Delhi, 2002, pp. 313–352.
86 K. R. Srinivasan, "The Megalithic Burials and Urn-Fields of South India in the Light of Tamil Literature and Tradition", *Ancient India*, ASI, New Delhi, 2, 1946, pp. 9–16.
87 S. B. Deo, "The Megaliths: Their Culture, Ecology, Economy and Technology", in Shantaram Bhalchandra Deo and K. Paddayya (eds.), *Recent Advances in Indian Archaeology: Proceedings of the Seminar Held in Poona in 1983*, Deccan College Post-graduate and Research Institute, Poona, 1985, pp. 89–99.
88 R. K. Mohanty and V. Selvakumar, *op. cit.*, pp. 313–352.
89 Udayaravi S. Moorti, *Megalithic Culture of South India: Socioeconomic Perspectives*, Ganga Kaveri Pub.House, Varanasi, 1994, p. xxiii, 134.
90 Elizabeth Chesley Baity and others, "Archaeoastronomy and Ethnoastronomy so Far [and Comments and Reply]", *Current Anthropology*, University of Chicago Press Journal, Vol. 14, No. 4, 1973, pp. 389–449.
91 A. Thom, "Megalithic Astronomy: Indications in Standing Stones", *Vistas in Astronomy*, Vol. 7, Elsevier, 1966, pp. 1–57.
92 David H. Kelley and Eugene F. Milone, *op. cit.*, p. 62.
93 Udayaravi S. Moorti, *op. cit.*, p. xxiii, 134 p; Udayaravi S. Moorti, "Megaliths", in Deborah M. Pearsall (ed.), *Encyclopaedia of Archaeology*, Elsevier, 2008, pp. 745–755.
94 A. Sundara, *The Early Chamber Tombs of South India: A Study of the Iron Age Megalithic Monuments of North Karnataka*, University Publishers (India), Delhi, 1975, p. 80.
95 Mayank N. Vahia and others, "Megaliths in Ancient India and Their Possible Association with Astronomy", in Tsuko Nakamura and others (eds.), *Mapping the Oriental Sky, Proceedings of the Seventh International Conference on Oriental Astronomy*, Tokyo National Astronomical Observatory of Japan, Tokyo, 2011, p. 15.
96 Srikumar M. Menon and Mayank N. Vahia, "Megalithic Astronomy in South India", in Tsuko Nakamura and others (eds.), *ibid.*, pp. 21–26 (p. 21).
97 Srikumar M. Menon, Mayank N. Vahia and Kailash Rao, "Stone Alignment with Solar and Other Sightlines in South India", *Current Science*, Vol. 102, No. 5, Bangalore, 2012, pp. 683–684.
98 K. P. Rao, "Astronomical Relationship of South Indian Megaliths", in W. Todd Bostwick and Bryan Bates (eds.), *Viewing the Sky through Past and Present Cultures*, Selected Papers from the Oxford VII International Conference on Archaeoastronomy, City of Phoenix Parks Recreation and Library, City of Phoenix, 2006, pp. 421–432.
99 K. P. Rao, "Ancient Sky-Map from Mudumal", *Man in India*, Vol. 91, No. 2, Serial Publishers, New Delhi, 2011, pp. 359–362.

100 *Ibid.*, p. 362.
101 *Ibid.*, p. 360.
102 N. Kameswara Rao, Priya Thakur and Y. Mallinathpur, "The Astronomical Significance of "Nilurallu", the Megalithic Stone Alignment at Murardoddi in Andhra Pradesh, India", *Journal of Astronomical History and Heritage*, Thailand, Vol. 14, No. 3, 2011, p. 211.
103 N. Kameswara Rao, "Astronomical Orientations of the Megalithic Stone Circles of Brahmagiri", *Bulletin of Astronomical Society of India*, Hyderabad, Vol. 21, No. 1, 1993, pp. 67–77.
104 Riza Abbas, Mayank Vahia, V. Shobha, R. K. Mohanty & Aniket Sule, *Search for Archaeoastronomical Data from the Megalithic Burials of Junapani, Nagpur, Maharashtra*, Tata Institute of Fundamental Research, 2010, p. 9; see also Abbas, "Cup Marks: Rock Art of Megalithic Stone Circles of Junapani", *Open Access Library Journal*, Vol. 1, 2014, pp. 1–6.
105 Uthara Suvrathan, "Landscapes of Life and Death: Considering the Region of Vidarbha", in Upinder Singh and Nayanjyot Lahiri (eds.), *Ancient India: New Research*, Oxford University Press, 2009, pp. 122–171.
106 *Ibid.*, p. 156.
107 Romila Thapar, *A History of India*, Vol. I, Penguin Books, 1965, Chapter-1.
108 Uthara Suvrathan, *op. cit.*, p. 126.
109 Mayank N. Vahia and others, *op. cit.*, p. 15.
110 *Ibid.*, p. 16.
111 *Ibid.*, p. 15.
112 N. Kameswara Rao, "Astronomy with Buddhist Stupas of Sanchi", *Bulletin of the Astronomical Society of India*, Hyderabad, Vol. 20, 1992, pp. 87–98.
113 F. R. Allchin, "The Stone Alignments of Southern Hyderabad", *Man*, RAI, UK, 1956, pp. 133–136.
114 *Ibid.*, p. 134.
115 N. Kameswara Rao and Priya Thakur, "The Astronomical Significance of Megalithic Stone Alignments at Vibhuthihalli in Northern Karnataka", *Journal of Astronomical History and Heritage*, Thailand, Vol. 13, No. 1, 2010, pp. 74–82; Priya Thakur, "Prehistoric Stone Alignments of Southern India", *CIJHAR*, M.P, Vol. 2, No. 6, 2013, pp. 1–6.
116 N. Kameswara Rao, Priya Thakur and Yogesh Mallinathpur, "The Astronomical Significance of 'Nilurallu', The Megalithic Stone Alignment at Murardoddi in Andhra Pradesh, India", *Journal of Astronomical History and Heritage*, Thailand, Vol. 14, No. 3, 2011, pp. 211–220 (p. 217).
117 K. Paddayya, "The Stone Alignment at Hanamsagar, District Gulbarga", in L. K. Srinivasan and S. Nagaraju (eds.), *Sri Nagabhinandanam*, Dr. M. S. Nagaraja Rao felicitation committee, 1995, pp. 23–28.
118 N. Kameswara Rao, *op. cit.*, pp. 499–511.
119 Srikumar M. Menon and Mayank N. Vahia, "The Monument in the Landscape: Using Remote Sensing to Understand the South Indian Megaliths", *International Journal of Scientific & Engineering Research*, Vol. 4, No. 12, 2013, pp. 192–199.
120 *Ibid.*
121 *Ibid.*, p. 21.
122 Srikumar M. Menon, *Cosmic Considerations in Megalithic Architecture: An Investigation into Possible Astronomical Intent in the Design and Layout of Megalithic Monuments with a View to Understanding Megalithic Knowledge Systems*, Manipal University, 2013, pp. 203–210.
123 *Ibid.*, p. 203.
124 *Ibid.*, p. 210.
125 *Ibid.*, p. 206.

126 Srikumar M. Menon, Mayank N. Vahia and Kailash Rao, *op. cit.*, pp. 683–684.
127 *Ibid.*, p. 683.
128 Clive L. N. Ruggles, "Megalithic Astronomical Sightlines: Current Reassessment and Future Directions", in D. C. Heggie (ed.), *Archaeoastronomy in the Old World*, Cambridge University Press, Cambridge, 1982, pp. 83–106 (p. 82).
129 D. C. Heggie, "Megalithic Astronomy: Highlights and Problems", in D. C. Heggie (ed.), *ibid.*, pp. 1–24 (p. S18).
130 Kelley and Milone, *op. cit.*, p. 162.
131 Juan Antonio Belmonte, "Ancient "Observatories - A Relevant Concept?", in Clive L. N. Ruggles (ed.), *Handbook of Archaeoastronomy and Ethnoastronomy*, *op. cit.*, pp. 133–45 (p. 142).

7 Sushruta: The Sage Surgeon of Ancient India and His Surgical Instruments

Amarnath Thakur

The golden age of surgery in ancient India rests largely on the accomplishments of Sushruta, the son of Sage Viswamitra and a disciple of Dhanvantari, Divodasa Kasi Raja, and the creator of *Sushruta Samhita*.[1] Sushruta's time has long been a controversial subject among many medical historians. Since the date has never been definitely set for lack of direct evidence, the scholars have placed the period, when he lived and practised, from 1000 BC to AD 1000. As many later Vedic hymns are ascribed to Sushruta, it follows that he must have flourished during the latter part of the Vedic age. Hoernle contended that a larger portion of *Atharvaveda* admittedly belongs to a period as early as 1000 BC since the hymns in question are included in the older portion.[2] However, Johnston-Saint suggests that Sushruta was a contemporary of Buddha because of the style of language used.[3]

Although taking the basis of the ancient form of construction of the Sanskrit language used in the manuscript, Wise has concluded that *Sushruta Samhita* was written in an extremely early age, probably from the 3rd to the 9th century BC.[4] Nagarjuna's *Upayahrdaya* (500 BC) refers to Sushruta, and this takes him to a period which was around 2000 years ago. Further, he is mentioned in both *Mahabhasya* of Patanjali and the *Varttika* of Katyayana. No other Sushruta is known except the medical writer who was the propagator of *salyatantra*. Incidentally, Hoernle has placed him at about 600 BC as he counts only 300 bones in the body, and on this ground, he believes him to have lived before Atreya and Yajnavalkya, thus taking him to somewhere around 600 BC[5] – a period which although more empirical than historical, which generally has been considered to be the same by the historians. Though, it is quite possible that his actual time predates it by many centuries.

As already stated, in the absence of definite evidence, it is the mythology which takes us through the history of medicine in India. Brahma, the creator of the universe, knew the basis of Ayurveda through meditation and passed it to Daksa Prajapati who, in turn, taught it to the Asvin brothers. Lord Indra learnt it from the Asvin brothers and passed it on to many sages, including Sage Bhardwaj (teacher of Atreya) and King Divodasa Kasi Raja (Lord Dhanvantari). Sushruta approached Dhanvantari to accept him as his disciple and learnt the basis of Ayurveda with special emphasis on *salya*

DOI: 10.4324/9781003370642-8

(Surgery). Later on, Sushruta composed *Sushruta Samhita* while teaching at Benares, situated by the river Ganges in North India.[6]

Samhita created an atmosphere of independent thinking and investigation which later characterized Greek medicine.[7] The original manuscript of *Sushruta Samhita* has not survived. Extant ones are only copies of copies and revisions of revisions. One of the most important sources in connection with ancient Indian medicine is the Bower Manuscript,[8] which is named after its discoverer Lieutenant H. Bower and is now housed in Oxford University Library. It was found in Kuchar in Eastern Turkestan in 1890 on the great caravan route of China. It was subsequently sent to Colonel J. Waterhouse, the then-President of the Asiatic Society of Bengal, where the famous Indologist, Hoernle, edited this document critically and placed its origin around the 6th century BC.[9]

Sushruta was one of the first to study human anatomy. In his *Samhita*, Sushruta has described in detail about the knowledge of anatomy with the aid of a dead body, and the techniques of dissection, enumerated therein, are unique and revealing.[10] This is the earliest record of human dissection undertaken in medical history for the study of anatomy and physiology. However, this antagonized the upholders of orthodoxy. Surgery was becoming synonymous with rationality, which became the most crucial factor for its development. Sushruta was the first person to advocate the practice of mock operation on inanimate objects and had devised various experimental modules for trying each procedure including watermelons, clay pots and reeds, thus predating the modern practice of the surgical workshop by half a millennium.[11] For example, incision and excision were practised on vegetables and leather bags filled with mud of different densities; scraping on the hairy skin of animals; puncturing on the veins of dead animals and lotus stalks; probing on moth-eaten wood or bamboo; scarification on wooden planks smeared with beeswax. Basham has possibly rightly observed that "India evolved a developed empirical surgery" and opined that "ancient Indian surgeons remained ahead of Europeans until the 18th century".[12]

Sushruta divided his *Samhita* into six parts, the *Purva-tantra* into five sections, viz. *Sutrasthana*, *Nidanasthana*, *Sarirasthana*, *Kalpasthana* and *Chikitsasthana* and the *Uttara-tantra*. These sections together encompass, apart from *salya* and *salakya*, the other specialties also like medicine, paediatrics, geriatrics, diseases of the ear, nose, throat and eye, toxicology, aphrodisiacs and psychiatry. They also cover all the branches of medicine including hygiene, midwifery, ophthalmology, toxicology, psychosomatic ailments and materia-medica. *Samhita* provides some 60 procedures for the management and healing of wounds. It also refers to the pathogenesis of the disease and the wounds in different stages and categories of surgical procedure. The whole procedure is divided into a three-phase exercise:

- *Poorva karma* (preparatory stage) – to prepare the patient, the operation theatre and the instruments

- *Pradhan karma* (principal act of operation) – surgical procedures, i.e. expulsion, stitching, medication, etc.
- *Paschata karma* (post-operative measures) – includes bandaging, redressing, cleaning the wounds, healing procedures, cosmetic restoration procedures, etc.[13]

These surgical procedures show a deep concern for the recovery of the patients which added a new dimension to *Sushruta Samhita*. For an extensive treatment of surgery, Sushruta has described the surgical methods in different heads,[14] such as excision (*chedana*), whereby a part or whole of the limb is cut off from the parent; incision (*bhedana*) is made to achieve effective drainage or exposure of underlying structures to let the content out; scrapping (*lekhana*) or scooping is carried out to remove a growth or flesh of an ulcer, tartar of teeth, etc. and the veins, hydrocele and ascitic fluid in the abdomen are drained by puncturing with special instruments (*vyadhana*); the sinus and cavities with foreign bodies are probed (*esana*) for establishing their size, site, number, shape, position, situation, etc.; blood-letting (*sravana*) is to be carried out in skin diseases, localized swellings, etc. and, in case of accidental injuries and intentional incisions, the lips of the wound were apposed and united by stitching (*svana*). The detailed instructions regarding suturing the wounds underline the level of Sushruta's surgical excellence.[15]

Sushruta describes 20 kinds of sharp instruments and 101 types of blunt instruments and their handling techniques. These instruments were further classified into four heads: *sastra* (cutting instruments), *yantra* (blunt instruments), *upayantra* (accessories appliance) and *anusastra* (minor instruments). These instruments are notable for their varieties and precision. Sushruta further adds that a surgeon, by his own experience and intelligence, may invent and add new instruments to facilitate surgical procedures. However, Sushruta held that the hand is the most important[16] and the best instrument, but for which the operation of other instruments ceases. The blunt instruments were meant for the removal of foreign bodies, for sucking the fluids, for facilitating the various surgical procedures and for visualizing the lesions. The double-armed axial instruments, which had two moving limbs to hold and pull any object, were called svastika-yantras and they resembled the various types of forceps.

Sushruta's classification and description became the basis for the development of instruments, such as specula, scalpels, scissors, saws, needles, cauterizes, syringes, trocars, catheters. These were usually given names according to their resemblances to animals, such as lion, cat, hawk, crocodile. In fact, Sushruta can be said to have been the first person to introduce diagnostic instruments and their principles, which were modified later with the introduction of an optical system in their construction.

Fourteen types of bandaging, capable of covering almost all the regions of the body, were described for the practice of the student on dummies.[17] Sushruta had covered the accidental burns in its four degrees, the effect of heat stroke, sunstroke, frostbite due to excessive cold and also the effect of lightning which he described as *vidyut-dagdha*. This classification underlined his view that all thermogenic traumas, whether due to extreme cold or heat, wet or dry, chemical or inert fluid, produced damage almost similar and hence were to be managed as one entity. This great value of Sushruta's classification could be realized from the fact that this concept gained validity in modern surgery only recently after 1950 and is now uniformly accepted in the classification and management of these injuries.

One pioneer contribution of Sushruta was in the field of orthopaedic surgery, wherein he gives categories of the bones and their reaction to injuries, varieties of dislocation of joints (*sandhimukta*) and fractures of the shaft (*khanda-bhagna*) in a very systematic way.[18] Sushruta gives the details of the 6 types of dislocations and 12 varieties of fractures, and also describes the treatment, thereof, by means of traction, manipulation, appositions and stabilizations. The most significant in this context is to develop the method of the use of a healthy limb to support an injured limb.[19] Moreover, Sushruta also had suggested exercises along with other techniques to treat limb injuries.[20]

Again Sushruta rendered, hithertofore, the unknown operation which later on came to be known as plastic surgery. He described various methods including sliding graft, rotation graft and pedicle graft for restoring several facial and bodily defects, such as rhinoplasty (repair of the nose), repair of cleft lips, otoplasty (repair of torn ear lobes). All the principles of plastic surgery, viz. planning, accuracy, precision, economy, haemostasis and perfection, find an important place in his writing on the subject. The method of rhinoplasty stood the test of time, and the technique has remained unchanged till today in all the books on plastic surgery.[21]

Sushruta was also the first surgeon to describe 76 different diseases of the eye of which 51 required operations. He was the first surgeon to have removed cataracts and to have described varieties of cataracts along with the depression method of couching by the anterior root.[22] However, it is interesting to note that while in Hindu medicine cataract was defined by Sushruta as opacity due to derangement of the intraocular fluid, subsequent history is full of fantasies and prejudices concerning its nature.[23]

In order to assuage or alleviate the pain of surgical incisions, Sushruta advocated the use of wine as an anaesthetic material.[24] Although the use of hemp (*Cannabis indica* Linn) and of *sammohini* and *sanjivani* is reported at a later period, Sushruta was, however, the pioneer of anaesthesia. Whipple opines that "all in all, Sushruta must be considered the greatest surgeon of the pre-medieval period".[25]

Influence of the Indian School of Medicine on Arab and Greek Medical Sciences

No discussion of Indian medicine is complete without discussing medicine in ancient Arab and Greece. It is interesting to note that several Indian tribes like Jats (Zutt), who had settled down in Arabia even before the beginning of Islam, were well versed in different branches of ancient Indian traditional medicine and were using their clinical proficiency to cure the patients. When the Arabs realized the high quality and value of Ayurveda, they got interested in the translation of Indian medical and other scientific works from Sanskrit into Arabic.[26] Thus *Sushruta Samhita* was also rendered into Arabic. The Arabic translation of these medical treatises highly impressed the Arabs who assimilated an enormous material in their Tibbi medical treatises.

Avicenna or Ibn Sina (AD 980–1037) was the most famous scholar of Arabic medicine who influenced the global medical literature for a long time. This 'Prince of Physicians' wrote the canon of medicine, *al-Quanun-fil-Tibb*, which has been used for centuries as the authoritative text on Unani medicine.[27] In this text, Ibn Sina expresses his indebtedness to the Indian doctors and quotes verbatim from Ayurvedic treatises. This book of Avicenna was translated into Latin in the 12th century AD and was a text-book in European medicine for long.[28] It may be concluded that the Arabs developed great respect and love for Indian medical scholars and their products as is evident from many historical references and panegyrics (*qasida*) composed in reply to the carping of a critic by an Arab of Indian origin, Abu Dila'Sindhi (9th century AD).[29]

However, the question of who borrowed medical knowledge from whom has always been a stimulating topic. A small school of medical historians believe that the East contributed nothing whatsoever to ancient medicine and that the Greek influence dominated medical practice in ancient India and other Arabic countries. Some even question the existence of Sushruta and his *Samhita*. Johann Hermann Bass, the German medical historian, went so far as to assert that Sushruta had been no other than Hippocrates. The name, he argued, had been confounded with Socrates in the Indian tenets and finally transmitted as 'Sushruta'. Bass saw proof of his thesis in the fact that Sushruta's birthplace was given as Kasi, an old name for Benares. That was obviously a distortion of the name of the Greek island of Kos, where Hippocrates was born.

Commenting upon this apparently distorted version, Gordon comments, "How one clever enough to read Hippocrates in the original and reproduce it in his own language could mistake the name of Socrates for Hippocrates surpasses one's imagination".[30] Gordon concludes that Indian medicine, especially its surgery, developed in ancient times most probably independent of Greek medicine. Thorwald, following his detailed discussion of Indian medicine, asserts that Greek medicine, including that variety of it called Hippocratian, had produced nothing in the field of surgery that could remotely compare with the striking ideas of Sushruta.[31] Royle, one of the earliest authorities of Indian medicine, gives evidence that Hippocrates borrowed much of his material medica from Indians.[32] Wise, another scholar in ancient Indology, says that it is to the Indians that we owe the first system of medicine.[33] Neuberger concluded that "Greek medicine adopted Indian medicaments and methods which are evident from the literature".[34] As to the relationship between Indian and Arabic medicine, it is well documented that works by Sushruta and Charaka were translated into Arabic.

It is reasonable to state that the ancient Hindu and Greek systems of medicine were independent of each other.[35] However, contact between the two civilizations first became intimate through the march of Alexander and continued unbroken through the reign of Diadochoi and the Roman and Byzantine eras. As a result, Indians probably borrowed some ideas from Greek medicine, while the Greek physicians of the time borrowed drugs and surgical techniques from ancient India.

Notes

1 M. Tewari, "The Father of Indian Surgery", *Indian Journal of Surgery*, Vol. 67, 2005, pp. 229–230.
2 A. F. R. Hoernle (ed.), *The Bower Manuscripts*, Aditya Prakashan, New Delhi, Vol. 1–3, 2011.
3 P. Johnston-Saint, "An Outline of the History of Medicine in India", *Indian Medical Record*, Vol. 49, 1929, p. 289.
4 T. A. Wise, *Commentary on the Hindu System of Medicine*, Thacker, Calcutta, 1845, p. 1.
5 A. F. R. Hoernle, *Studies in the Medicine of Ancient India Part I: Osteology or the Bones of the Human Body*, Clarendon Press, Oxford Press, Oxford, 1907, p. 109.
6 F. McDowell, *The Source Book of Plastic Surgery*, William & Wilkins Co., Baltimore, MD, 1977, pp. 65–85.
7 N. H. Keshwani (ed.), *The Science of Medicine and Physiological Conceptions in Ancient and Medieval India*, Dept. of Physiology, All-India Institute of Medical Sciences, New Delhi, 1974, p. 16.
8 A. F. R. Hoernle (ed.), *op. cit.*, Vol. 1–3.
9 V. K. Baskaradoss and R. M. Jose, "Life and Times of Sushruta: Fact and Fiction", *Indian Journal of Plastic Surgery*, Vol. 39, 2006, pp. 109–10.
10 K. K. Bhisagratna, *The Sushruta Samhita* (English translation based on the original Sanskrit texts), *Sutrasthana* (Henceforth *SS*), Chowkhamba Sanskrit Series Office, Varanasi, India, 1963, 5, pp. 49–56.
11 *Ibid.*, *SS*, *Sutrasthana*, 8.3, 8.12, 12.8, 25.3, 25.7; *Chikitsasthana*, 22.15–16, 22.18–20.
12 A. L. Basham, *The Wonder that was India*, Fontana, 1971, p. 501.
13 *SS*, *op. cit.*, *Sutrasthana*, 5.1–2.
14 *Ibid.*, *SS*, *Sutrasthana*, 5.3–15; 7.12; 8.3.
15 G. N. Mukhopadhyaya, *Ancient Hindu Surgery: Surgical Instruments of the Hindus*, Cosmo Publications (Reprint), New Delhi, 1994. Vol. I.
16 *SS*, *op. cit.*, *Chikitsasthana*, 2.46.
17 *Ibid.*, *SS*, *Sutrasthana*, 2.5, 2.2–3, 25.43–44.
18 *Ibid.*, *SS*, *Nidanasthana*, 3.15–16.
19 *Ibid.*, *SS*, *Chikitsasthana*, 3.37–40; 3.47–48.
20 G. D. Singhal, S. N. Tripathy and G. N. Chaturvedi (trs.), *Fundamentals and Plastic Considerations in Ancient Indian Surgery*, Singhal Publications, Varanasi, 1981, p. 315.
21 A. N. Sharma, "Surgical Techniques", in A. K. Bag (ed.), *History of Science & Technology*, Vol. I, Indian National Science Academy (INSA), New Delhi, 1997, p. 342.

22 V. K. Raju, "Susruta of Ancient India", *Indian Journal of Ophthalmology*, Vol. 51, 2003, pp. 119–122.
23 W. S. Duke-Elder (ed.), *Systems of Ophthalmology*, CV Mosby, St Louis, MO, 1969, Vol. XI, p. 65.
24 *SS*, *op. cit.*, *Chikitsasthana*, 7.2, 15.7, 16.20–23; *Nidanasthana*, 8.10.
25 A. O. Whipple, *The Story of Wound Healing Repair*, Charles C. Thomas, Springfield, IL, 1963, p. 18.
26 M. Z. Siddiqi, *Studies in Persian and Arabic Medical Literature*, Calcutta University, Calcutta, 1959, pp. 31–43.
27 R. L. Verma, "Indian-Arab Relations in Medical Sciences", in P. V. Sharma (ed.), *History of Medicine in India*, Indian National Science Academy (INSA), New Delhi, 1992, pp. 465–484.
28 C. Elgood, *A Medical History of Persia*, Cambridge University Press, Cambridge, 1951, pp. 278–301.
29 D. P. Agrawal and Lalit Tiwari, *Ayurveda: The Traditional Indian Medicine System and its Global Dissemination*, The Infinity Foundation (On the Internet).
30 B. L. Gordon, *Medicine throughout Antiquity*, A. Davis Co., Philadelphia, PA, 1949, p. 323.
31 J. Thorwald, *Science and Secrets of Early Medicine* (Translated by R. Winston and C. Winston), Harcourt, Brace & World Inc., New York, 1963, p. 194.
32 J. F. Royle, *An Essay on the Antiquity of Hindoo Medicine*, William H. Allen & Co., London, 1837.
33 T. A. Wise, *Commentary on the Hindu System of Medicine*, Trubner & Co., London, 1860, pp. i–iv.
34 M. Neuberger, *History of Medicine* (Translated by E. Playfair), Oxford University Press (OUP), Vol. I, 1910, p. 60.
35 U. Prakash, *Shushruta of Ancient India*, Sur Gyn and Obs, 1978, 146:263.

8 Traditional Knowledge of Medicine in Ancient India

(*Based on Classical Texts and Treatises*)

Seema Sahay

Since time immemorial, human beings have always aspired to achieve a good and sound state of health and well-being. The very concept of being healthy, with body and mind, has always been there. Like every other thing, diseases and afflictions along with their cures and remedies, too, caught human attention for ages. The very idea of being sick and ailing was not liked even by those living in ancient times. So, it is believed that people in the ancient era made efforts to understand the phenomena that caused afflictions so that they could be competent and proficient enough to eliminate their harmful effects. In the course of time, such efforts felicitated the rise of the art of healing in ancient India that would ensure not only prevention and cure of diseases and ailments, but prolongation and maintenance of healthy life too.

India has a rich history of the traditional system of medicine that proudly illustrates the attempts made to understand the ancient art of healing. There are various literary sources, classical treatises and documentaries that give a broad and comprehensive picture of the evolution, growth and development of ancient Indian medical lore. Such classical sources need to be examined so that we may adopt a more judicious approach towards understanding the practice of medicine in ancient India.

The history, growth and development of classical Indian medicine can be discussed under two distinct phases – the Vedic era and the post-Vedic era.

If we talk about the Vedic age, we find that there is no one or more than one specific compendium that is entirely devoted to the evolution, growth and development of ancient Indian medicine. There are several Vedic scriptures, i.e. the Vedas, the Brahmanas and the Upanishads that give us some useful information on the concept of medicine or the art of healing of the period. With the help of various references scattered here and there, we can create a fairly logical and adequate picture of medicine in the Vedic era.

To trace the evolution, growth and development of the art of healing in the Vedic times, we largely depend on the Vedas, which are religious texts that originated in ancient India. The Vedic age has four Vedas; out of these four Vedas, the *Rigveda* and the *Atharvaveda* are considered to be the principal sources of the history of medicine. The textual evidence of ancient Indian medicine is largely found in the *Atharvaveda*, and as far

DOI: 10.4324/9781003370642-9

as the *Rigveda* is concerned, it throws significant but limited light on the aforesaid subject. We come across several such hymns in these two Vedic scriptures that help a lot in constructing the history, growth, development, custom and practice of the ancient system of medicine.

In the post-Vedic age, it seems that a more systematic, methodical and logical system of medicine developed. The two most renowned treatises of the period were *Charaka Samhita* and *Sushruta Samhita*. These two distinguished classical treatises were composed by two sages: Charaka and Sushruta, respectively. Though the *Samhita* of Sushruta is primarily a book on surgery, *Charaka Samhita* is considered to be the compendium of therapeutic medicine that gives an in-depth knowledge of the study of diseases and their remedies in ancient times.

A careful examination of such classical treatises will help us to further our understanding of how the whole system of medicine was in play; how the diseases and their symptoms and diagnosis were perceived in ancient India. It has also been found that the above-mentioned classical treatises, i.e. the *Rigveda*, *Atharvaveda*, *Charaka Samhita* and *Sushruta Samhita* have several references to the various kinds of diseases that were prevalent during the period; though it could be said that the nature and characteristic of the ancient system of medicine varied from time to time. For instance, it seems that the perception of diseases and their treatment during the Vedic age was more magico-religious in character.[1] While in the post-Vedic era, the later medical traditions and practices were more rational, systematic and scientific in their approach rather than relying on the magical and religious justifications of diseases and remedies.[2]

To begin with, we will first get acquainted with the art of healing that was prevalent during the *Rigveda* and the *Atharvaveda*. Both the classical texts are considered to be the earliest of the sacred scriptures that would together help to create a framework that would explain the medical tradition of the period.

But before we embark on it, here it would be interesting to trace the origin of ancient Indian medicine which is largely believed to be of divine origin. There is a mythological story that narrates how the knowledge of the science of medicine was passed from heavenly personalities to sages and rishis. According to the story, Brahma was the first to propose the art of healing through Daksha Prajapati who later taught this science to Aswin Kumars. Aswin Kumars then imparted the science of the art of healing to Indra. It is believed that up to this time, the science of medicine was only known to divines like Brahma, Aswins and Indra, and it was Indra who transmitted the science of life to mortals like sages and rishis, Bharadwaj being his first pupil.[3] Girindranath Mukhopadhyaya, in his book *History of Indian Medicine*, has also mentioned Brahma to be "the propounder of the healing art – Ayurveda or the 'Science of Life', which is said to be an *Upaveda* of *Rigveda*".[4]

Even during the Vedic age, divinity played a vital role in the medical tradition of both the Rigvedic and the Atharvavedic times. Primarily hailed as a "liturgical book of the *hotrs*", *Rigveda* is a collection of hymns that are mostly devoted to various divinities who largely represent the Vedic pantheons.[5] Since it has already been suggested that the *Rigveda* is mostly religious in character, it has been opined by Kenneth G. Zysk that "it provides for the most part mythological stories illustrating the healings performed by various gods of the Vedic pantheon".[6] Hence, taking his viewpoint into consideration, it is principally noticed that most of the hymns in the *Rigveda* are on healing deities like Aswins, Rudra, Maruts, Bhaskar.

In the *Rigveda*, there are several hymns devoted to Aswins. The Aswins were known as the divine physicians who had remedies against all diseases.[7] They are said to have restored youth to the old and injured body of the sage Chyavana.[8] In one such story, they are credited to have restored the eyesight of Rijrasava and provided to Vispala an artificial leg that had been cut off in a battle.[9]

There are also numerous references in the *Rigveda* that extol Rudra, Maruts, Indra and Bhaskar also as the healers who brought about cures and remedies. In one such verse of the *Rigveda*, Rudra is praised as "the best of physicians", "the depository of all sciences" and "the possessor of healing medicines".[10] Lord Indra is said to have possessed healing powers that could cure any kind of wounds.[11] It is even suggested that Bhaskar, the Sun, who is the father of the Aswins, is a good physician who dispels diseases.[12] In one such hymn, he is invoked to dispel the disease of the heart and the yellow colour of the body – "Oh! Sun, cure the sickness of my heart and yellow colour of my body".[13]

So, all such hymns or verses mentioned in the sacred scripture further corroborate the fact that there were several deities who were often engaged in healing activities. They were all supposed to be the possessor of medicines, healing powers, remedies and cures that would bestow good health and well-being. All such hymns also give glimpses of the disease that afflicted the people. They also give us useful insight into those factors that caused the diseases and the kind of treatment that was employed to get rid of them. Some of the diseases mentioned in the *Rigveda* are *yaksma*, *rapas*, *amiva*, etc.

In the *Rigveda*, *yaksma* is described as the kind of disease that enters and possesses each and every part of the body.[14] There is yet another hymn or charm in the *Rigveda* that deals with the removal of *rapas*. With regard to *rapas*, R. Muller has suggested that the word *rapas* was "a general expression for fragility and sickness".[15] In the *Rigveda*, the disease *rapas* has been explained as a "bodily affliction that often attacked the foot".[16] Another hymn shares a similar view and says that *rapas* is used to describe "a sick person who seems to have been suffering from crooked or deformed limbs".[17] *Amiva* is another such disease or affliction that has been frequently referred to in both the *Rigveda* and in the *Atharvaveda*.[18] The numerous references in the *Rigveda* have described the term *amiva* as the "domestic demon" and has described the disease as something which is related to lack of nourishment.[19]

As we go further in our quest to know about the kinds of diseases prevalent in the *Rigveda*, we come across one common factor. That common factor was based on one fundamental belief that diseases and afflictions were caused by the demonic entities that entered the body and caused sickness and ill health. These demons or evils that caused the diseases were often sent by a god as a punishment for the sin they committed. It was mostly the evil spirits, demons and other malevolent forces that attacked the victim's bodies and caused a state of being sick and diseased.[20] For example, in the *Rigveda*, the disease *rapas* has been described to have a "divine origin" and it is also associated with various evils and demons that caused diseases.[21] One such evil mentioned is *ajakava* that has been described as an "evil-looking and crawling animal". It was believed that it resided under the skin and caused eruptions and swellings in the joint.[22] A similar view has been expressed by Thomas T. Mackie who has also described various kinds of medical conditions to be of demonic origin. He further elaborates that the demon is a worm-like creature that attacks the body of the victim from water and eventually poisons him.[23] As described earlier as a domestic demon, *amiva* was often envisaged as an evil that came with a "host of other evils or demonic elements" that punished the body for the sins committed.[24]

So, based on the above discussion, one aspect that seems to be apparent so far is that the diseases and maladies were either of demonic or divine origin. It was mainly believed that diseases were sent by gods for the sins they committed or were caused by evils and demons that entered the body, which eventually led to sickness and indisposition. Now, since they owed the causes of diseases and ailments to Vedic divinities and demons, they often engaged in healing activities, rites and rituals that would usually please the Vedic pantheons. By pleasing gods and divinity, they would ensure a healthy body and mind. This one aspect could be better explained with the help of several instances.

In the case of *yaksma*, there are verses in the *Rigveda* that show oblations were paid to Indra and Agni to drive away the demons that had possessed the body of the ailing and get rid of the malady which caused distress and miseries among the sick and ailing.[25] In the treatment used for the removal of *rapas*, often the Vedic gods like Mitra, Varun and Agni were invoked to protect and guard them against the disease.[26] Another form of remedy applied for the removal of *rapas* was water or watery medicine.[27] Another disease, *amiva*, was also dealt with a similar approach like any other disease. In the *Rigveda*, various divinities were called upon to remove *amiva*. Divinities like Aswins, Agni, Soma, Rudra were beckoned and pleaded to chase away the sickness.[28] In one of the hymns, Indra is said to have been invoked to help the worshippers overcome penury and hunger caused by *amiva* and give food and drink to the hungry.[29]

So, based on our discussion pertaining to the Rigvedic times, several interesting aspects have come into the picture that in general explain the nature and characteristics of the system of medicine that prevailed in the Rigvedic age. As far as the perception of diseases and their causes and treatments are concerned, one thing is clear that the whole approach towards medicine was no doubt religious in character as highlighted by scholars like P. Kutumbiah and Kenneth G. Zysk. P. Kutumbiah opined that the diseases were caused by the "possession of evil spirits, anger of certain gods, by evil deeds, and the sorcery of enemies".[30] Perhaps this draws attention to one interesting characteristic that there was no rational or scientific explanation for whatever malady, affliction or disease plagued them. As far as the cures and remedies of the disease were concerned, it was usually seen that since the diseases were owed to demons, evil spirits and gods, the mode of treatment also followed the same line of reasoning, devoid of any scientific explanation and pragmatic approach. People believed that if they make the god happy and do not commit any sin, the god will be pleased with them and will bestow upon them joy, health and strength. Gods will ensure them with cures and remedies and will also nurse them back to good health. This theory about the perception of diseases and their treatment seems to be believable as we do not find any such hymns that demonstrate that a more rational and scientific approach was adopted towards the system of medicine that was in vogue during this phase.

Another important aspect that comes to our mind is whether the Rigvedic healers made an effort to understand the symptoms of diseases. It is still difficult to fathom whether they took symptoms into consideration before diagnosing diseases. It seems that there is no such definite evidence to show otherwise. Though, we have noticed in the previous paragraphs that in one of the hymns, the Sun God or Bhaskar was invoked to cure the sickness of the heart and "yellow colour of the body".[31] Now, here one could only speculate what the yellow colour of the body meant. Was this a kind of disease or a symptom of a disease? Were the people in the Rigvedic era aware that diseases afflicted the body while symptoms were the indications of those particular diseases which helped them to understand the ailments better? Taking the religious character of the medicine into consideration and looking at all the mythological stories that surround the Vedic pantheon and their healing activities, it is hard to believe otherwise, but if a part of it is true, then this empirical approach towards the art of healing surely comes as a surprise.

From the *Rigveda*, we come down to the *Atharvaveda* – an ancient text that has verses and prose usually chanted for the performance of magical rites by priests, atharvans and healers. The sacred scripture is replete with verses and prose that are usually in the form of magical charms, spells and incantations. Such verses and prose are mostly uttered for various purposes like acquiring sharp intelligence, owning riches, winning battle, and protecting oneself from the rage and fury of enemy.[32] Besides the charms that were usually recited for general purposes, there were also charms that were used for medical cures and remedies for a variety of diseases like fever, cholera, epilepsy, heart disease, jaundice, leprosy, consumption, dropsy, diseases of the head, ear, nose, throat, eyes and neck, and possession of a person by a demon or an evil spirit. These are some of the diseases that are commonly talked about in the *Atharvaveda*.[33]

When we look at the exhaustive list of the assortment of purposes for which the charms were used and uttered, we notice that charms, spells and incantations formed a significant part of the life of people of the time. With the help of charms and spells, the people not only aspired to fulfil their social needs but also made it an essential part of their art of healing technique. The same idea has been corroborated by O. P. Jaggi, who rightly pointed out that

> the list of diseases not only gives us the glimpse of the mode of thinking, their social needs, behaviour and environment of the people of that time, but it also throws light on the art of healing of the time.[34]

If we take the medical practices, customs and beliefs of the Vedic period into consideration, we find that besides the *Rigveda*, the *Atharvaveda* also emerges as a vital, interesting and informative source that gives us an abundance of insight into the medical lore of the time. Like in the *Rigveda*, people during the time of the *Atharvaveda* also had mythical conceptions about diseases and their causes and cures. Even during that time, it was a common belief that diseases had a divine or demonic origin. According to popular belief, there were several factors that led to poor and failing health. Committing sins in the present or past life, disrespect shown to gods and witchcraft of the enemies were some of the common factors that were considered to have caused diseases and maladies. It was also believed that god, in alliance with various demons, made people suffer for the sins they committed and for any kind of misconduct done by them.[35] Similar ideas about the mythical conception of diseases have also been expressed by Zysk who opined that "diseases were believed to be produced by demonic or malevolent forces when they attacked and entered the bodies of the victim".[36]

Based on the above views, ideas and insights, one could say that diseases and maladies were the manifestations of wrath and fury of numerous gods and demons brought upon those who committed sins or displeased gods in one way or another. For instance, Varun, the cloud god, was responsible for causing many diseases, and one of the diseases attributed to him is dropsy. This particular disease caused the whole body to swell up.[37] Rudra was another god who was said to have inflicted illness and acute pain. It was said that Rudra with his shooting arrows and missiles caused fever, cough and acute pain.[38] Besides Varun and Rudra, Takman was another god who caused fever, and in one such verse, he is being appeased to banish fever.[39]

Besides the diseases mentioned above, there were some other diseases that were said to have been caused by demonic entities. Some of the chief ones were *yaksma*, *ksetriya*, *rapas*, *balasa*, *amiva*, ascites, insanity, *krimi*, urine retention, etc.[40] Other than the above-mentioned diseases and ailments, there were different forms of afflictions like broken bones and limbs, flesh wounds and fractures.[41] When a careful glance is given at the names of various diseases found in the *Atharvaveda*, one thing that comes into the picture is that based on the causative factors of the diseases, people had developed some kind of understanding of the classification of diseases.

It was rightly pointed out by Zysk that there were maladies that affected the inside of the body that were caused by disease demons, and there were wounds, injuries, broken bones and other similar afflictions that resulted from accidents, battles and pests as well.[42] Perhaps this understanding of diseases is our first peak into the idea that there was some sort of classification of diseases done during the Vedic times. It also indicates the very fact that there were diseases which caused internal afflictions, and there were diseases that caused external afflictions. Writing about the concept of diseases, S. N. Dasgupta holds the view that there was a three-fold classification of all diseases known to the Atharvavedic people: diseases produced by wind, those produced by water and those that are produced by fire. According to him, these fundamentals about the classification of disease are close to the later classification of diseases that was based on three dosas (wind or *vayu*, phlegm or *kapha* and bile or *pitta*).[43] So, based on the above view, it could be inferred that the concept of *Tridosa* that is so deeply engrained in Ayurvedic medicine could be traced back to the *Atharvaveda*. One could only wonder how far this theory or view holds true, but it seems to be reasonable to accept that there was surely some idea about the classification of diseases in the *Atharvaveda*. Such classifications were usually done on the basis of those causes and factors that led to diseases.

As far as the mode of treatment was concerned, it was found that the whole fundamentals of the treatment of diseases were directly associated with the concept of the origin of diseases or the causative factors of diseases. It has been already mentioned that the general attitude towards the cause of diseases and infirmities was mythical and magico-religious in which the origin of diseases and maladies was owed to factors like evil spirits, demons and malevolent forces. Based on these assumptions, the kind of treatment employed during this time to bring a cure was also largely magical based. Various healing techniques and practices were adopted and applied to drive away maladies, illnesses and sicknesses of all kinds. It was found that often divinities were beseeched by the use of incantations, and a wide variety of charms and spells were uttered to ensure sound health. So, in general, it was found that magic, charms, spells and amulets together were used to ward off evils and neutralize the effect of evil and demonic forces. Besides these, medicinal herbs, plants and their concoctions and the use of water as medicines were also a significant part of the mode of treatment.[44]

In the *Atharvaveda*, charms and incantations were held in special regard. The incantation of magical verses and utterance of charms played a vital role in bringing about the cure and warding off the harmful effects of demons and evils. The charms and spells were usually a part of a healing ritual where often healers also known as *bhisaj* recited the required magical verses to drive away maladies and banish all kinds of illnesses.[45] For instance, for treating diseases like fever, diarrhoea, diabetes, sores, leprosy, bleeding, jaundice, gods were propitiated and charms were recited for the removal of diseases.[46] The list for the usage of charms and spells for the removal of disease and ailments is endless. But one thing is certain that they surely played a significant role in the contemporary medical tradition.

Another aspect of the mode of treatment that was in trend in the *Atharvaveda* was the use of plants and plant products. It has been seen that medicinal plants and herbs were widely used in the practice of medicine and formed a noteworthy part of the healing ritual. There is one particular hymn in the *Atharvaveda* which is chiefly devoted to the virtues of medicinal herbs. This particular hymn gives us a valuable insight into the whole procedure of how the different herbs were brought into play for treating diseases during the Vedic period. In the hymn, there are detailed descriptions of different plants and herbs that have healing virtues and power to facilitate all possible help for all kinds of diseases. The hymns basically implied that all plants and plant products had healing powers. They all possessed specific characteristics and virtues of their own, and based on their beneficial properties, they were used accordingly to bring about the cure.[47]

Commenting upon the importance of plants and herbs in the Vedic medical tradition, Zysk says that during this phase, plants were treated as a goddess, and hence, they were venerated and respected for the healing virtues they possessed. He further elaborated that the description of a variety of vegetation in several of the medical hymns signifies that the people were well versed in the knowledge of flora and vegetation around them. Moreover, they also had gained an understanding of the "plant taxonomy" even though the appreciation of its efficacies was basic and elementary.[48]

Besides one such hymn discussed above, there are several other hymns in the *Atharvaveda* that provide helpful information about the usage and efficacies of different herbs. One such healing plant goddess was Arundhati, and the other one was Kustha. Arundhati was used in the treatment of wounds and fractures.[49] The healing plant Kustha was considered to be an important medicine for the cure of fever. Kustha, the healing god, was considered to be a "divine aromatic plant with all pervading strength and the medicine for all diseases".[50] There is also a mention of *cipudru* and *haridra* plants which were used for curing *balasa* and leprosy, respectively.[51] Wide varieties of plants and plant products were also used to treat other diseases like eye diseases, diseases of the mind, for increasing the growth of hair, food poisoning.[52]

Besides the use of charms and medicinal herbs, amulets, water and ointment (*anjana*) also formed a significant part of the healing technique applicable in the Vedic time. With regard to the use of water as watery medicine, there is one specific hymn in the *Atharvaveda* which is devoted to water and its healing properties.[53] Water and watery medicines were used for the removal of *rapas*.[54] Amulets were also considered to be potent and were usually used together with plants and plant products. They were further made more effective by the use of magical verses. In one such hymn, it is mentioned that ointment was used to cure jaundice and *jangida* amulet was employed to protect the patient from *balasa* and *takman*.[55]

From our thorough discussion made above, one thing that seems to be evident is that the healers had various options and approaches to cure the kind of maladies that affected the health of the people. To bring about a cure and a complete state of well-being, the healer could make use of charms or spells, medicinal herbs, amulets, etc. It looks like that all such techniques either worked independently on their own or sometimes worked in conjunction with each other. For instance, as mentioned above, amulets were usually used in combination with plants and plant products and were often made more potent and effective with the utterance of magical verses or charms. Here, one could only imagine that the healers used their experience to follow a particular mode of treatment to bring about effective results.

So far, it is clear that there are many hymns in the *Atharvaveda* that throw an abundance of light on the different kinds of diseases that were prevalent and the mode of treatment that was employed to eradicate the same. All such hymns also shed light on the characteristic of Vedic medicine. It seems probable that two types of art of healing or system of medicine existed: one was based on magic which predominantly made use of magical spells, charms and incantations, and the other was the use of different plants and their products which were used in combination with the utterance of magical verses and charms. It was believed that there were priest physicians who mostly relied on magical practices like the recitation of magical verses and charms to bring about a state of good health. But there were also healers or medical practitioners who were well versed in the effectiveness of plants and plant products and they used herbal medicines in a more rational manner to eliminate diseases. The extensive use of medicinal herbs during this time probably marks the beginning of a more rational and empirical mode of thought and understanding about diseases and their cures.[56]

A similar idea has been expressed by Zysk who elucidated that in due course the Vedic healers developed a more "systematic and classificatory mode of thinking"[57] and eventually became "familiar with more empirical procedures of healing".[58] He further elaborated that as far as diagnosis of diseases and detailed knowledge of plants and their efficacies were concerned, Vedic healers were capable of observing and identifying symptoms in order to recognize diseases and study the benefits, virtues, qualities and valuable facts of potent plants. Perhaps they observed facts and made records of persistent or recurring symptoms and various characteristics of patients for future reference. The same approach was adopted while studying the importance and role of plants and herbs in the earliest medical system. In fact, "the technique of recording valuable facts is an example of the very beginning of the Indians empirical mode of thought".[59]

From the *Rigveda* and the *Atharvaveda*, we now come down to an age when the system of medicine seemed to have developed more on an empirical and rational mode of thought. It looks like that during this time, the study of medicine and its application were more methodical and systematic and were largely followed with a more scientific and pragmatic approach. As far as the treatises of this period are concerned, some of the earliest and renowned treatises of this period are *Charaka Samhita* and *Sushruta Samhita*. *Charaka Samhita* is primarily a book on medicine, while *Samhita* of Sushruta largely deals with surgery.

Talking about the nature and characteristics of medicine, it was witnessed that in the *Atharvaveda*, there were two modes of treatment that were usually employed and applied for the removal of diseases and afflictions. One mode of treatment was based on the utterance of charm, which was designated as magico-religious medicine. Another mode of treatment was based on medicinal herbs that were hailed as empirico-rational medicine. Based on the theory put across by P. Kutumbiah, with the passage of time and owing to several factors, the two systems of medicine became independent of each other. It seems that in course of time the system of medicine based on magic and religion lost its relevance and a more empirical and rational system of medicine gained prominence.[60] We may agree with the views of Zysk that "magical medicine did not completely vanish from traditional Indian medical lore, it was never to acquire the status in the Ayurvedic treatises which it had enjoyed in the *Atharvaveda*".[61]

It seems that the shift from magico-religious to empirico-rational continued till the ancient system of medicine became more logical, methodical and systematic and was no longer religious in character. Perhaps it was this system that laid the foundation of early Ayurveda, also known as the "Science of Life", that provided a more reasonable and scientific explanation of diseases and their cures. However, as rightly commented by P. Kutumbiah, "Both Charaka and Sushruta describe Ayurveda as an *Upanga* or an *Upaveda* connected with the *Atharvaveda* and acknowledge it as the source of their medical knowledge".[62] Based on the above discussions, it seems to be clear that a part of the medical tradition of the *Atharvaveda* continued even during the post-Vedic era that seemed to have influenced the later medical texts.

Talking about the later medical texts, according to Zimmer, "*Charaka Samhita* stands as the finest document of the creative period of ancient Indian medicine".[63] It is a medical text that has discussed in detail "the science of the causes and symptoms of diseases, of their medicine and other maintenance of health".[64] The ancient text consists of a detailed account of the classification of diseases, causative factors of diseases, their diagnosis and their treatment as well. Besides dealing with various diseases and their cures, it also emphasizes personal hygiene and sanitation and the training of physicians.[65] As already mentioned earlier, *Charaka Samhita* is a classical treatise that is largely devoted to therapeutic medicine and deals with its various aspects. Though, it also deals with other branches like anatomy (*sarira*), physiology (*vritti*), aetiology (*hetu*), pathology (*vyadhi*), treatment (*karma*).[66] The treatise is divided into eight divisions which in general deals with topics like principles and philosophy of medicine, causes, diagnosis, prognosis and treatment of diseases, nourishment, anatomy, embryology, pharmacy.[67]

Such a systematic classification and division of topics concerning health and medicine clearly point out that an earnest effort had been made to make the medical tradition more methodical and orderly. It seems that at this juncture, the whole approach towards medicine was more practical, logical and scientific. For instance, good and sound health could not be only maintained by just eliminating diseases, but good hygiene and sanitation also played an important role in staying healthy.[68] Similar ideas related to personal hygiene and sanitation have also been highlighted in other classical texts. In *Sushruta Samhita*, washing hands, feet and mouth before eating has been suggested.[69] In another such treatise, Manusmriti, it has been advised that "without a reason don't touch your own indriyas, i.e. eyes, nose, ears, mouth, etc."[70] Coming to the present day, we find that these very suggestions and ideas seem to be similar to the thoughts and beliefs of what is being practised and followed nowadays. In the current pandemic situation, which has gripped the entire world, good hygiene and sanitation are considered to be crucial to reduce the spread of infection. Several cleaning and hygiene tips and measures have been recommended to stay safe from the virus. Moreover, training physicians were considered to be vital as it is done today. In *Charaka Samhita*, itself, the practice of medicine by "unqualified persons or self-styled physicians"[71] was highly criticized and disapproved of. All these aspects point out that any matter that concerned health was taken seriously. They are also indicative of the fact that during this time a methodical and systematic mode of treatment was considered to be more important that was largely based on scientific precision. The following discussion will further support the aforesaid theory.

Charaka Samhita gives us a detailed description of a wide variety of diseases. The names of the diseases mentioned in the text are endless, and it is not possible to include them all here. Some of the common diseases that have their nomenclatures in *Samhita* are related to the eyes, nose, throat and ears. Then there are also diseases related to the heart, stomach, intestines, jaundice, leprosy, urinary disorders, skin diseases, etc. The text also mentions some common diseases like dysentery, diarrhoea, baldness, asthma, diabetes.[72] Besides the various diseases mentioned in the treatise, there is also a methodical description of pathological conditions of diseases. For instance, the disease *padasula* has been described as a disease that causes pain in the foot, and *supti* is a disease that causes cramps or temporary paralysis.[73] Likewise, there are innumerable diseases with their pathological conditions mentioned in the aforesaid text.

The detailed knowledge of various diseases with pathological conditions emphasizes that physicians of the time tried to acquire a systematic knowledge of the essential nature of disease that would further help them to understand the causes of diseases and their cures and remedies as well. Such well-thought and methodical approaches towards the understanding of diseases indicate that the system of medicine was becoming more rational and scientific. Contrary to the Vedic belief and perception, the physicians did not rely on any magical or supernatural explanation for the occurrence of any disease; rather they tried to look for the real cause that would help them to tackle the diseases in a skilful manner.

As to the concept of the origin of diseases, Charaka maintained that the *Tridosa* or the humoral theory played a crucial role in the causation of diseases and their pathology, diagnosis and classification and treatment as well.[74] The humoral theory is based on the idea that there are three humor or dosas in the human body: *vayu* (*vata*), *pitta* and *kapha*. Based on this concept, Charaka believed that any disharmony or imbalance of the three dosas makes people vulnerable to diseases.[75] It was further elaborated that if dosas are in equilibrium, then a person is considered to be healthy, but if any one of the dosas predominates, the person is said to be susceptible to diseases.[76] Besides recognizing humoral theory as the cause of any occurrence of diseases, Charaka believed that the possibility of diseases also arose due to external factors like environment, climate, water supply and seasonal variations. For instance, he tried to explain that those living in humid climates and dense forests had delicate physiques and health, while areas with a proper balance of dry and moist seasons contribute to good health.[77]

Unlike the Vedic healers, the cause and origin of the diseases were not considered to be demonic and divine in nature. They in fact took an empirical trajectory for the explanation of those factors that lead to sickness and ill health. Owing to the origin of diseases to humoral theory and various other external factors seems to be more logical and technical in nature. The very belief that permutations and combinations of the three dosas in varying degrees may lead to numerous pathological conditions seems to be more believable and realistic rather than owing the cause of diseases to demons, gods and unnatural forces. It is even seen that before treating the disease, the physicians were also asked to find out which *dosa* was not in balance or displayed abnormal nature, quality and characteristics.[78]

As we move further to understand the system of medicine that prevailed in the post-Vedic era, it is surprising to find that *Samhita* also contains a general theory and methods of diagnosis. If we go back to the Vedas, the uncertainty about any method or theory with regard to the diagnosis of diseases still persists. But coming down to this time, we see for certain that the physicians took diagnosis into consideration based on the symptoms and prognosis of diseases.[79] According to *Samhita*, there were three ways for physicians to discover and determine the nature of diseases. First was the theoretical knowledge of all the causes and symptoms of diseases.[80] The second way was to carefully and thoroughly observe the patient's symptoms and complaints, and the third way was the inferences that were largely based on previous experience.[81]

Looking carefully at the ways and means of diagnosis done during this period, one could only wonder how similar the whole approach is to the one done today. Speaking of the present age, doctors and physicians, in this day and age, take a similar approach prescribed in the classical text. They, too, take the patient's symptoms and complaints into consideration. Then after careful examination and observation, and based on their previous experience and knowledge of the disease, they draw the inference. Here, one could say that the knowledge and experience of the physicians mattered a lot. Even in *Samhita*, the role of the ancient physicians in diagnosing diseases was highly emphasized. It has been mentioned that the physician will be unable to reach any conclusions if he has incorrect knowledge of the disease. The ancient physicians were expected to have a detailed and profound knowledge of the nature of all diseases with regard to their origin, indications, symptoms and possible complications. Besides the role of physicians, the patient's history and background also played a crucial role in the diagnosis.[82]

As to the knowledge of the classification of diseases, it seems that a proper understanding of the nature and characteristics of various diseases had developed. Perhaps it was this understanding that finally helped to make smooth progress towards classifying diseases under various heads based on their origin, symptoms and the parts of the body they afflicted. According to *Charaka Samhita*, diseases were categorized as curable and incurable and mild or violent. Classification of diseases was also done on the basis of the difference in their causes. For example, there were diseases that were caused due to constituent elements, and then there were diseases that owed their origin to accidental causes. Similarly, diseases were also grouped on the basis of where they arose from in the body. For instance, there were diseases that arose from the *amasaya* (stomach) and there were diseases that arose from the *pakvasaya* (intestines).[83]

Once the causative factors of the diseases were ascertained and a diagnosis made, the next course of action was the treatment of diseases. There is an expansive range of treatments for diseases mentioned in *Charaka Samhita*. According to *Samhita*, all substances whether living or non-living are considered to be medicines only if they are applied in a correct manner (*yukti*) and for a specific reason (*artha*).[84] Based on this theory, medicines, cures and remedies were derived from various sources like plants and plant products, animals and the earth. Honey, fat, marrow, skin, blood, flesh, etc. were animal products used as drugs. Gold, silver, sand, lime, salt, etc.

were some of the drugs derived from the earth.[85] The drugs derived from plants and plant products were root, bark, fruit, flower, oil, etc. All such products were considered as drugs and medicines that were used for the treatment of all kinds of maladies. Medicines derived from all such products were used and applied in different ways.[86]The use and application of medicine were usually done internally or externally. For example, honey is prescribed to include in the diet for all weak conditions, while in other forms of diseases it is used externally for local applications.[87] Medicines derived from a large array of sources were prescribed and administered in different forms; sometimes, they were used as ointments, paste, fumigants and oils and sometimes they were included in diets in the forms of drinks, mixtures and decoctions.[88]

Besides prescribing various medicines for the correction of diseases and ailments, *Samhita* also recommends several prophylactic measures for the maintenance of health, vigour and vitality. Such prophylactic measures included certain courses of action that were required to be a part of an individual's daily routine. For instance, regulated physical exercise was highly recommended for good digestion, energy and strength.[89] A good, healthy, nutritious and wholesome diet was also greatly emphasized for being hale and hearty. It is mentioned in *Samhita* that "a wholesome diet is stated to promote bodily growth and health; unwholesome food is the most important cause of diseases".[90] If we happen to see the mode of treatment employed these days, we find that even today regulated exercise and a proper diet in health and sickness are highly stressed upon and advocated by modern doctors and physicians.

Based on our discussion, one could say that as compared to the Vedic age, the medical tradition of the later period was more empirical and a more judicious, sensible and well-thought approach was adopted towards the understanding of diseases and ailments. Such an appreciation went a long way in developing and establishing a system of medicine that took a more scientific, realistic and holistic approach in dealing with all the matters and concerns of health.

Notes

1 Kenneth G. Zysk, *Medicine in the Veda: Religious Healing in the Veda*, Motilal Banarsidass, Delhi, 1998, p. 7.
2 *Ibid.*, p. 11.
3 O. P. Jaggi, *Indian System of Medicine*, Atma Ram & Sons, Delhi, 1981, C.S.1.1. pp. 1–2.
4 Girindranath Mukhopadhyaya, *History of Indian Medicine* (Vol. I), Oriental Books Corporation, Reprint, New Delhi, 1974, p. 172.
5 Kenneth G. Zysk, *Asceticism and Healing in Ancient India: Medicine in the Buddhist Monastery*, Motilal Banarsidass, Delhi, Reprint, 2000, p. 14.
6 Kenneth G. Zysk, *Medicine in the Veda, op. cit.*, p. 5.
7 O. P. Jaggi, *Dawn of the Indian Medicine*, Atma Ram & Sons, Delhi, 1969, pp. 166–167; *Rigveda* (Henceforth RV) RV VIII 18.8, 22.10; for Rigvedic hymns,

see R. T. H. Griffith (trans.), *The Hymns of Rigveda*, The Chowkhamba Sanskrit Series Office, Varanasi, Reprint, 1971.
8 O. P. Jaggi, *Dawn of the Indian Science*, *ibid*., p. 166; RV I.116.10.
9 *Ibid*., p. 167; RV I.112.8.
10 See also Girindranath Mukhopadhyaya, *History of Indian Medicine* (Vol. I), *op. cit*., p. 172 (Introduction); RV 2.7.16.
11 *Ibid*., p. 100 (Chap-1); RV vii.i, v.12.
12 *Ibid*., p. 82; RV I.35.9.
13 *Ibid*., p. 82; RV.1.50.11–13.
14 Kenneth G. Zysk, *Medicine in the Veda*, *op. cit*., p. 12; RV 1.122.9; 10.97.12; 10.161.5; 10.163.
15 *Ibid*., p. 25.
16 *Ibid*., RV 7.50.1–3.
17 *Ibid*., see reference no. 14 in p. 25.
18 *Ibid*., p. 49.
19 *Ibid*., RV 7.71.2; 8.48.11; 10.37.4 and 6.74.2.
20 *Ibid*., p. 8.
21 *Ibid*., p. 25; RV 2.33.7; 1.34.11; 1.57.4.
22 *Ibid*., p. 26; RV 7.50.1–2.
23 Thomas T. Mackie, et al., *A Manual of Tropical Medicine*, W. B. Saunders Company, Philadelphia, PA, 1945, pp. 362–366.
24 Kenneth G. Zysk, *Medicine in the Veda*, *op. cit*., p. 49; RV 1.35.9–10 and 1.189.1.
25 Girindranath Mukhopadhyaya, *History of Indian Medicine* (Vol. I), *op. cit*., p. 106; RV 10.161.1.
26 Kenneth G. Zysk, *Medicine in the Veda*, *op. cit*., p. 25; RV 7.50.1–3.
27 *Ibid*., p. 26; RV 7.50.4 and 8.20.23–25.
28 *Ibid*., p. 50.
29 *Ibid*., pp. 50–51; RV 10.42.10 and 1.104.7.
30 *Ibid*., pp. 5, 8; P. Kutumbiah, *Ancient Indian Medicine*, Orient Longmans, Hyderabad, 1962, p. xi.
31 Girindranath Mukhopadhyaya, *History of Indian Medicine* (Vol. I), *op. cit*., p. 82; RV.1.50.11–13.
32 O. P. Jaggi, *Dawn of the Indian Medicine*, *op. cit*., pp. 168–169.
33 *Ibid*., p. 169.
34 *Ibid*.
35 *Ibid*., p. 170.
36 Kenneth G. Zysk, *Asceticism and Healing in Ancient India: Medicine in the Buddhist Monastery*, *op. cit*., p. 15.
37 O. P. Jaggi, *Dawn of the Indian Medicine*, *op. cit*., pp. 170–171; AV II. 9.3; for Atharvavedic hymns, see William Dwight Whitney (tr.), Charles Rockwell Lanman (ed.), *Atharvaveda Samhita*, 2 Parts (Henceforth AV), Motilal Banarsidass Publishers Pvt. Ltd., New Delhi, 1971.
38 O. P. Jaggi, *Dawn of the Indian Medicine*, *op. cit*., pp. 170–171; AV VI.90.
39 *Ibid*., p. 172; AV I.25.
40 For the diseases mentioned, see Kenneth G. Zysk, *Medicine in the Veda*, *op. cit*., pp. 12–70; AV 2.33, 6.85; 2.8, 3.7; 4.13, 6.91; 6.14, 6.127; 7.42(43).1–2; 4.16; 6.111; 2.31; 1.3.
41 *Ibid*., p. 72; AV 4.12; 5.5.
42 Kenneth G. Zysk, *Asceticism and Healing in Ancient India: Medicine in the Buddhist Monastery*, *op. cit*., p. 15.
43 S. N. Dasgupta, *History of Indian Philosophy*, Cambridge University Press, 1922, Vol. III, p. 300.
44 O. P. Jaggi, *Dawn of the Indian Medicine*, *op. cit*., pp. 173–174.

45 Kenneth G. Zysk, *Medicine in the Veda*, *op. cit.*, p. 8.
46 O. P. Jaggi, *Dawn of the Indian Medicine*, *op. cit.*, p. 174; AV I.37, VII.116; I.23; I.171; I.22.
47 *Ibid.*, pp. 203–206; AV VIII.7; Also see the translation by H. R. Zimmer, L. Edelstein (ed.), *Hindu Medicine*, The Johns Hopkins Press, Baltimore, MD, 1948, p. 15.
48 Kenneth G. Zysk, *Medicine in the Veda*, *op. cit.*, pp. 99–102.
49 Kenneth G. Zysk, *Asceticism and Healing in Ancient India: Medicine in the Buddhist Monastery*, *op. cit.*, p. 18.
50 Kenneth G. Zysk, *Medicine in the Veda*, *op. cit.*, p. 39; AV5.4.1,2; 19.39; 5.22.3; 19.39.5–9.
51 *Ibid.*, p. 32; AV 6.127.2; I.24,2; O. P. Jaggi, *Dawn of the Indian Medicine*, *op. cit.*, p. 211.
52 O. P. Jaggi, *Dawn of the Indian Medicine*, *ibid.*, pp. 178, 212.
53 Kenneth G. Zysk, *Medicine in the Veda*, *op. cit.*, pp. 90–92; AV 6.24.
54 *Ibid.*, p. 26; AV 6.91.3; AV 6.57.1–3.
55 For the cure of Jaundice, see O. P. Jaggi, *Dawn of the Indian Medicine*, *ibid.*, pp. 175–176; AV II.28; for the cure of *balasa* and *takman*, see Kenneth G. Zysk, *Medicine in the Veda*, *op. cit.*, p. 33; AV 19.34.10.
56 P. Kutumbiah, *op. cit.*, pp. xiv–xv.
57 Kenneth G. Zysk, *Medicine in the Veda*, *op. cit.*, pp. 9–10.
58 *Ibid.*
59 *Ibid.*
60 P. Kutumbiah, *op. cit.*, p. xvii.
61 Kenneth G. Zysk, *Medicine in the Veda*, *op. cit.*, p. 11.
62 P. Kutumbiah, *op. cit.*, p. xvii.
63 H. R. Zimmer and L. Edelstein (eds.), *op. cit.*, pp. 52–53.
64 P. Ray and Hirendra Nath Gupta, *Caraka Samhita: A Scientific Synopsis*, The National Institute of Sciences of India, New Delhi, 1965, p. 4; Su.1,23.
65 *Ibid.*, p. 4.
66 *Ibid.*, p. 4; Si.12, 53; Su.30, 32.
67 *Ibid.*, p. 5; Su.30, 35.
68 *Ibid.*, p. 4.
69 "*Naprkhalitpanipado bhuyajit*" (meaning: "Wash your hands, feet, mouth before you eat".), *Sushruta Samhita*, Chikitsa Sthan, 24.98.
70 "*Anaturah swani khani na sprisheda-nimittatah*" (meaning: "Without a reason don't touch your own indriyas, i.e. eyes, nose, ears, mouth, etc."), *Manusmriti*, 4.144.
71 P. Ray and Hirendra Nath Gupta, *op. cit.*, pp. 4, 21; Su. Chap.15.
72 *Ibid.*, pp. 99–107 (see the tables that display the names of diseases with their modern names).
73 *Ibid.*, pp. 110–111; Su.20,11; Ci.27,16.
74 *Ibid.*, p. 10; Su, 7, 40.
75 *Ibid.*
76 O. P. Jaggi, *Indian System of Medicine*, *op. cit.*, p. 116; CS, 1.3.7; 37.8.
77 P. Ray and Hirendra Nath Gupta, *op. cit.*, pp. 17–18; Vi.3,4; Ka.1,8; Vi.3,7,47.
78 *Ibid.*, p. 11; Ni.1,13.
79 *Ibid.*, p. 21.
80 *Ibid.*, p. 21; Vi.4, 3–5.
81 *Ibid.*; O. P. Jaggi, *Indian System of Medicine*, *op. cit.*, p. 124.
82 P. Ray and Hirendra Nath Gupta, *ibid.*, pp. 21–22; Vi.4, 6–9; Vi. 8,93.
83 O. P. Jaggi, *Indian System of Medicine*, *op. cit.*, p. 121; CS. 3.6, 1–3.
84 P. Kutumbiah, *op. cit.*, p. 110; CS I.25,27.
85 *Ibid.*, pp. 110–111; CS I.1.

86 *Ibid.*
87 P. Ray and Hirendra Nath Gupta, *op. cit.*, pp. 43–44; Ci.26, 253; Su.5,43.
88 *Ibid.*, pp. 38–84.
89 *Ibid.*, p. 12; Su.7, 31–35.
90 *Ibid.*, p. 18; Su.25, 31.

9 Indic Sciences in Tibet

The Development of Ayurveda in Tibet and Its Transformation as Tibetan Medicine

Melarkode Narayanaswamy Rajesh

This chapter tries to locate the development of Tibetan medicine from its Indic roots and the transformations from Ayurveda to a hybrid system incorporating elements from Tibetan folk medicine and other influences from China and other regions. All of these were synthesized under Buddhist influence in the institutional base of the Buddhist monastery. Thereafter, Tibetan medicine acquired a distinct personality with a large collection of texts and developed practices.

Modern history writing is characterized by a shared universality of methods which is reflected in a set of attributes that define what history is, and the three main characteristics are verifiability, secular causation and a sense of anachronism.[1] Though there are many different methodological approaches that vary from one end of the spectrum to another, encompassing diverse approaches, all of them hold certain aspects as necessary. They are all dependent on the primary sources constituted in the form of a hard archive that can be verified, contested and reinterpreted within a particular domain of possibility.[2] Such practices of history, therefore, possess a universal currency and can be translated throughout the modern world and are played in differing contexts, as long as the hard archive is available to be interpreted in Africa, Asia or Europe. However, in the case of Africa, we see that the hard archive limits and distorts the sources as the narrative of history in Africa is primarily oral and translating it into a print is problematic as it would rob the meanings.[3]

This was the situation even a century ago in some contexts where there were differing practices of history that did not follow the written word and also relied on many other forms of tradition, as in case of pre-colonial South India before the dislocation of traditional historical narratives and the entrenched march of the modern historical narratives in the wake of the colonial encounter.[4] Therefore, to understand Tibetan medicine from a textual point of view is very problematic as it is closely intertwined with religion.

Just as capitalism is a world system and can be translated universally in a wide variety of contexts throughout the capitalist world as it has penetrated almost all corners of the globe, pre-capitalist systems like feudalism

DOI: 10.4324/9781003370642-10

are notoriously hard to define for the simple reason that they are situated in a wide variety of contexts that are rooted in regional particularities. Just as the presence of the European church and its attendant public do not square off and find a parallel even with the mediaeval Ethiopian Orthodox Church as a feudal institution or the samurai of mediaeval Japan who was characterized variously as feudal but does not find a one-to-one correspondence with the feudal institutions and the categories of people who are part of feudalism, we have a problem of correspondence of categories. Thus, a mediaeval European knight cannot be equated with the samurai of Japan. Even in the case of history writing, such parallels are easily found, and this is the problem with pre-modern historical writing. Thus, the ideas of religion that became predominant in Tibetan medicine cannot be divorced from the narrative of medicine as a science.

Among the many characteristics of historical writing in a pre-modern world, there were wide variations not only in terms of style and form but also in terms of content. What actually comprised the content of historical writing was vastly different in many parts of the world. Thus, the griots of Africa[5], who are the traditional custodians of history in West Africa, had narratives that deferred from the Kula Puranas or the caste histories of the different castes of South India[6] or from the shamanic histories of inner Asia. Therefore, two points become clear when we start discussing the nature of orality: one that pre-modern history writing exhibited a vast range of diversity, and these could not be translated from one to one and, therefore, are very context specific. In contrast, the modern forms of historical writing presuppose a sense of universality which also reduces many aspects and the ambit of history to certain narrow domains.

In the modern period, most of the historical material that has come down to us from non-western societies is often approached and analyzed through western frames of reference, and here, once again there is the search for the hard archive. On the contrary, the source materials are to be read in a different context and an attempt to read them through an empirical framework would only reduce the scope of the material concerned. In the case of Tibetan history, and more particularly in the case of Tibetan medicine, this point is important since Tibetan medicine is understood to be part of the Tibetan astro and medical sciences, and therefore, ignoring the astrological and astronomical element would not give us a clear picture of the context of Tibetan medicine.

Like many pre-modern societies, Tibet also exhibited many of the characteristics of history writing that combined both the written and the oral tradition and the performative, and in the case of medicine, it was the performance of rituals. As Tibet had a 'Great Tradition' modelled in part largely from the Indian Sanskritic tradition and also various little traditions, the 'Great Tradition' provided a unifying reference point. In this sense, Tibet stood away from many of the neighbouring societies of Central Asia as Tibet also had a great tradition, whereas many parts of Central Asia did not have a script or a state. In this aspect of literarization, Tibet borrowed much from India and this process of literarization were largely based on the Indic Buddhist model derived from the great monasteries of mediaeval India.[7] Therefore, a large element of Indian scientific tradition was also translated and significantly adapted to the Tibetan context. In traditional Tibet, the lamas were the literary elite and, therefore, composed many texts including historical texts and in this process imparted a Buddhist tilt to most of the texts as they were devout Buddhists. In addition to religious histories, Tibetan literature also included a significant body of medical texts that contributed to the development of the Tibetan medical system.[8]

Astronomy and astrology were essential for understanding medicinal theory and practice, and since they were all deeply intertwined, it is not possible to understand them as independent disciplines. Further, both astronomy and astrology drew from Indian models with some Tibetan and Chinese influences and, therefore, they are called Indic science. As the above systems of knowledge developed in India and catered to all religious traditions of ancient India, these are labelled as Indic sciences.[9] In the recent period, there are certain issues in Tibetan historiography that seek to question the role and extent of Indian influence in Tibet. One of the arguments is that as the *lama* historians were the literary elite, therefore their narrative of history was naturally imbued with a bias towards Buddhism and India and, thus, exaggerated the role of Indic influence at the cost of Tibetan achievements. These arguments emerged in the 1990s and have sought to downplay the role of Indian influence.[10]

However, in the realm of medicine, we are on firmer ground as the materia medica of Tibetan medicine is predominantly sourced from Tibet itself. Further, Tibetan medicine caters to diseases of the Tibetan population which are also an outcome of their life in the high altitude eco-regions of Tibet rather than in the tropical and subtropical regions of India, and one class of Tibetan literature on the substitution of ingredients called *tsab* (*tshabs*) lists around 8000 formulas and also gives a list of substitution of ingredients.[11] Most of the diseases that are prevalent in India are virtually unknown in Tibet, and an example would be that of malaria as there were no mosquitoes in Tibet, and hence, malaria was unknown in Tibet.[12]

In the case of Tibetan medicine, one can be sure about the presence of the Indic sciences as it is easily comprehended in the materia medica, humoral system and reference to the texts. In some areas of Indic knowledge like iconography, Indian influence could be ignored or bypassed in some aspects like butter sculpture. In the case of the butter sculpture, which is one of Tibet's peculiar contributions, we do not have any adherence to Indic models.[13] Therefore, this comparison shows that there are spaces where Indic models are predominant and places where there is no mention of Indic models like butter sculpture. Nevertheless, these domains are not rooted in the everyday requirements of the topography and human adaptation to the topography as in the case of medicine. Thus, the iconography changed in Tibet from the period of the pre-Buddhist Bon religion to the emergence of Buddhism and its near total erasure in 1959 after the rise of the Communist onslaught on Tibet. On the other hand, the common diseases that haunted the people on the Tibetan plateau remain pretty much the same. This, therefore, explains the limited role of Indic influence in Tibetan medicine and argues that, as a historiographic exercise, one need not over-emphasize the role of Indian influence in Tibetan tradition.

A historical exercise to trace the origins of *So-ba-rig pa* or the Tibetan system of medicine would reveal two clear influences that are interconnected – the Indian influence and the Buddhist influence as medicine came in a big way to Tibet with the arrival of Buddhism in the 7th century AD. This period signalled the rise of a Tibetan script based on the *Gaudi* script that was the prevalent script in eastern India and in the great monasteries like Nalanda that became one of the prime sources of Tibetan Buddhist tradition. As the development of medicine took many twists and turns in India and its practice in the institutional framework of monasteries spanned more than 1000 years, there were many stages of development in India which led to the evolution of a model. In the case of Tibet, however, they did not have to go through this period of evolution as a ready-made model was already available to them from the great monastic universities or the Mahaviharas of eastern India. Therefore, the organizational basis of the Buddhist religion influences the course of the development of Tibetan medicine right from the period of the Buddha. In this narrative of history of medicine, many of the popular stories are taken from the life of the Buddha and the early Buddhist period and retold in Tibet and also find their resonances in the Tibetan tradition. One of the most important personalities here is the physician Jivaka. There are many accounts of Jivaka in the early Pali tradition and which are also found in the Tibetan tradition, and he is credited with having cured many diseases.[14]

The other important personalities of Ayurveda like Charaka and Sushruta are also mentioned in the Indian tradition, both in the mainstream Sanskrit and the Pali traditions and also in the later Buddhist traditions in most parts of the Buddhist world including Tibet. One of the major differences, here, regarding the origins is the Buddhist orientation given to these narratives buys replacing Brahma the creator god with the Buddha as the source of all medical knowledge.[15] In Tibetan Buddhism, where there is a pantheon of deities, the medicine Buddha is called Sangye Menla and is depicted in blue colour. One may note that while Ayurveda also affected other Buddhist societies like Sri Lanka that followed the Pali Buddhist tradition, the names of historical personalities like Charaka, Sushruta and Jivaka find a place, but we do not have any parallel to the medicine Buddha of Tibet.[16] Tibetan Buddhism has a large pantheon of divinities drawn from the Indian Tantric tradition and also from the indigenous Tibetan tradition, and thus, it allowed for a large number of deities, and there was room for a new deity associated with medicine – the medicine Buddha or Sangye

Menla. Thus, Tantric Buddhism, with its institutional base and the Tibetan context, mediated the arrival of Ayurveda in Tibet and, thus, conditioned its reception here.

The Tibetan system of medicine is known as *So-ba-rig-pa* and is a well-formed medical system like many other traditional medical systems.[17] However, there are many medical systems like Australian bush medicine and African medicine that claim to be independent medical systems, but they are predominantly a folk medical system. *So-ba-rig-pa*, on the other hand, is a fully developed system as it has a well-developed corpus of medical literature, an institutional base, a long historicity and fully articulated practice, and above all, detailed syllabi and a training programme of the doctors. These aspects have enabled *So-ba-rig-pa* to become part of the modern medical tradition by engaging with modernity as the products could be standardized. Keeping all these aspects in mind, the Indian government has currently accepted *So-ba-rig-pa* as one of the six systems of medicine that are officially patronized in India and are known by the nomenclature AYUSH.[18]

The first two letters of this acronym represent Ayurveda and the third letter 'U' represents the Unani tradition of medicine which came to India from the Middle East and is largely associated with Muslim culture and the Persian and Arabic languages. The letter 'S' represents both the *Siddha* medicine and tradition of Tamil Nadu and *So-ba-rig-pa*. The *Siddha* tradition dates back from the early historic period and processes a corpus of literature in one of India's classical languages, Tamil. It is also centred in Tamilagam and the northern parts of Sri Lanka which is again an integral part of the Tamil culture area since a very ancient period. Finally, the letter 'H' represents homoeopathy and is the only medical tradition, though alternate, which does not have a long historical standing in India.

So-ba-rig-pa is included as part of the Indian medical tradition because of the fact that there are traditional Tibetan culture areas in India, both in the western and eastern Himalayas, ranging from Ladakh, Lahaul, Spiti and Kinnaur in the western Himalayas to Sikkim in the central Himalayas and Tawang in the eastern Himalayas. All these regions are part of the Tibetan culture area and have close links with the neighbours like Nepal, Bhutan and of course Tibet, and more particularly central Tibet, where the great monasteries were located and which are also the source of all knowledge including medicine and Indic sciences.[19]

The development of Tibetan medicine was singularly possible because the monastic network led to knowledge production, conservation and transmission over the centuries all over the Tibetan culture area.[20] While the monks or the lamas were the literary elite in pre-modern Tibet, a few of them specialized in the study of medicine, whereas some others specialized in Buddhist philosophy or in particular religious and theological aspects of Buddhism including Buddhist logic and languages that attracted the bulk of students and thus flocked to the largest monasteries. The monastic network spanned over a very large area extending from Ladakh in the southwest to the borders of northeastern Tibet in the east and Mongolia in the northernmost point including all of Tibet in the centre. One of the peculiarities of the monastic system is the network of monasteries that fostered specialization and helped in finding a place for different monks who had special interests in certain subjects from far and wide.

Undoubtedly, the largest monasteries were situated in Lhasa and in the vicinity were monks from almost all parts of the Tibetan culture area gathered. To accommodate this diversity, the monks were housed in separate hostels, and therefore, there were separate hostels for monks from different places. There are many examples of monks flocking to the great monastic universities of Lhasa for higher studies from all corners of the Tibetan culture area and of the oft-quoted example is that of Agvan Dorjieff who came all the way from Buryatia in Russia to study in Lhasa in the 20th century.[21] We also have examples from the southernmost part of the Tibetan culture, i.e. the western Himalayas of India and two monks, namely Lama Kachan Lobsang Zopa, who was born in Ladakh in 1922 and went to study in the famous Tashilunpo monastery. Following him, another monk Sempa Dorji (born 1933) from Spiti also went to Tashilunpo monastery for higher studies.[22]

Many monasteries also specialized in certain aspects of knowledge and craft that made them renowned as centres specializing in one area like the famous monastery of Narthang that was known for printing. Similarly, the Gyuto monastery of Serra was well known for specific Tantric practices. It is, therefore, in this context of the Tibetan monastic environment that one has to understand the growth of medicine as religion, and the monastic element was predominant in all aspects of Tibetan life. Unlike other societies, in Tibetan society, a quarter of the able-bodied population according to Tibetan estimates and a slightly lower figure of around 18%, which is nearly one-fifth of the population according to Chinese estimates, withdrew from normal life and entered the monastery to become monks for a lifetime.[23] Thus, the monastic institution, which is primarily a religious element, contributed in a very large measure to the shaping of the Tibetan culture and mind.

Songtsen Gampo, the first Emperor of Tibet, is remembered at large for his military contributions and uniting the warrior nation of Tibet and extending the military might of Tibet to various parts of Central Asia, China and Nepal. Nevertheless, Songtsen Gampo is also remembered for bringing Chinese culture to Tibet which has become a rhetorical game between the Sino-Tibetan narratives of history. Transcending this aspect, we find that there are very important people in his court like the Indian physician Bharadwaja, Han Wan Heng the Chinese and Galen, who is identified as a person from Turkistan, which in all probability meant the west.[24] All of them met in a debate that was known as the great debate in Samye under the patronage of King Trisong Detsen, and this episode is taught to the students even today and shows the interaction of the different medical systems in Tibet.[25] While this may not have been that real Galen of the west, it could be someone who was part of the Galenic system of medicine, and therefore, it emerges that Tibet had access to many knowledge systems like Indian Ayurveda, Chinese medicine, which was predominant in East Asia, and the equally popular Galenic medicine of the west, in addition to the indigenous Bon medicinal tradition.

While the Chinese system had the humors – the yin and the yang – symbolizing the male and female elements, which in medicinal terms is understood as hot and cold. According to the Galenic system of medicine, there are four humors that are yellow bile, black bile, phlegm and blood disorders, and this particular categorization of humoral elements is the same as the classical Greek theory of humors. Incidentally, the Bon tradition also shared aspects of the Galenic system indicating a clear westward orientation. In fact, Tonpa Sherab Miwo, the founder of the Bon religion who occupies a place similar to the Buddha in the Bon religion, visited Tibet only once from the western country, and he found that Tibet was not yet ready to receive the teachings of the Bon.[26]

There is considerable debate on the location of the sacred geography of the Bon religion, and according to early cartographic representations in the indigenous tradition, it is to be located somewhere on the borderlands of Persia. This would then explain the idea of western elements in the Bon religion and the prevalence of ideas from the western regions. The Bonpo narratives refer to this region lying to the west of Tibet as Olmo Lungring and also as Olmo Tazig. The term Tazig would lead us to speculate whether this is roughly the Tajik region of the modern period. This fact assumes importance as India had regular contact with this region and the people of this region were represented in Indian history from a very early date and are known by the ethnonym as Tajika in Sanskrit sources. One of the important deities of the Bon religion, Okar, the deity of luminosity who becomes the white Tara in Tibetan Buddhism, is said to have originated from the Persian goddess Astarte for its source can be traced to the Sumerian goddess Ishtar. This explanation is given to emphasize the point that the borderlands of western Tibet served as a receptacle for influences from Persia, Greece and other parts of inner Asia.[27]

There is not much clarity on the Tibetan folk tradition and its humoral theory, but one can clearly say that the prevalence of these three systems of medicine shows that Tibet was in a position to choose whichever type of medicine it wanted as it was an empire and a dominant power and any knowledge system could not be imposed on Tibet. This also goes against the grain of Chinese claims of Songtsen Gampo's marriage as a civilizing encounter[28] where Chinese elements were adopted by Tibet.[29] In this case, it is neither the medicine nor the script that is adopted from China or the western regions but from India.

With regard to the mythological element, one can also talk about the Tibetan *Ramayana*, which is a clear example of Indian influence in Tibet during the early period. Dhondup Gyal was one of the modern Tibetan writers who spearheaded the movement to reclaim ancient Tibet's past by getting rid of the superimposition of any Indian influence on early Tibetan literature. His rationale was that the early Tibetan *Ramayana* was not a direct superimposition of Indian influence but exhibited a process of selectivity, wherein ancient Tibetan had a lot of power and was a warrior nation and, thus, had the option of selecting whichever influences they wanted. Dhondup Gyal, the person who commented on this aspect, was one of the leading poets of modern Tibet and is canonized on both sides of the border. While we see that in the case of the Tibetan *Ramayana*, the basic story structure, like many other Ramayanas all over southeast Asia and Central Asia, followed the process of certain changes like giving a Buddhist flavour by paying obeisance to the arahats and other changes like the depiction of Sita as Ravana's daughter.[30] As we had outlined in the case of medicine too, Brahma's place in Indian Ayurveda was substituted by Buddha in the Buddhist narratives of the Ayurveda tradition, and such a narrative kept all other parts intact and only substituted some of the main parts.

Some of the landmark features of Tibetan medicine are the main text, *rgyud-bzhi*, which is the translation of the Sanskrit text *Ashtangahridayasamhita* and contains around 5000 verses in 156 chapters organized in 4 volumes. The text *rgyud-bzhi* was rewritten many times and there are different versions, but the present version referred to was supposed to have been translated by the legendary physician Yuthok Yonthan Gonpo, the younger. Like the composition of religious literature, we find that the same monastic premises, which served as a space for the translation of religious literature from the Indic tradition to Tibetan, was employed for medicinal translation also as there was a well-established tradition of translation. Translation was carried out by a specialized set of monks known as the *lotswas* who were proficient in both Tibetan and Sanskrit and many times it involved a Tibetan monk and an Indian collaborator.[31] The system of translation, therefore, provided a reliable method for translating the various aspects of a disease, ideas of the body and related aspects.

In Tibet, *rgyud-bzhi* became the most important medical text.[32] This text attained great popularity and was also adopted in Mongolia[33] and places wherever Tibetan Buddhism spread. Thus, we have a Mongolian *rgyud-bzhi*, a Buryat *rgyud-bzhi* and so on. A strong monastic base and a textual tradition helped in the growth and spread of Tibetan medicine, and thus, we can see its efficacy in producing *amchis* or doctors who are in high demand even today. One of the examples of the popularity of the Buryat *rgyud-bzhi* is the number of doctors hailing from this region who became very popular in Saint Petersburg during the time of Czar Nicholas II. With the help of these doctors like Tsybikov, Badmaev and others, a thriving medicinal practice was established in Saint Petersburg which also led to the founding of the first Buddhist temple in Saint Petersburg helped by the growing clientele of Buryat doctors.[34] This was the northernmost frontier of Tibetan medicine where it interacted with the alternative medical practices of Europe in the 20th century.

In addition to the monastic base, the Tibetan medical tradition was also carried on by lineages of doctors, and the most important of them was Yuthok Yonthan Gonpo. As a contemporary of the Dharma King, Trisong Detsen, he said to have participated in the medical conference held under the patronage of the same king at Samye, the first Buddhist temple of Tibet.[35] As the early propagation of Buddhism seems to be problematic in Tibet due to the opposition of the indigenous deities, Acharya Padmasambhava from India was specifically invited to subdue these dangerous deities as the earlier *Acharya* from Nalanda, named Santaraksita, failed. Whatever was built during the day was destroyed in a night in the place of the first temple called Samye under the aegis of Santaraksita. Therefore, Padmasambhava, a famed Tantric master from the Vikramashila monastery, was invited. His method was not to subdue or eliminate the indigenous deities who destroyed the Buddhist temple but to accommodate them, and this new strategy paid off. Therefore, a compromise was affected and the deities of the earlier Bon religion were accommodated into Tibetan Buddhism as guardian deities.[36]

This had its resonance in medicine also where many of the indigenous deities are associated with certain healing practices, and one can notice that the propagation of Buddhism did not eliminate the earlier deities, but only brought about a change in configuration leading to the emergence of a hybrid religious tradition. This hybrid religious tradition grew from the indigenous Tibetan tradition represented by the Bon religion, the Tantric Buddhist tradition of eastern India and aspects of the Tibetan folk tradition. A Tibetan folk tradition conditioned the received understandings in such a way that the regional variations of Tibetan religion incorporated folk deities. Similarly, in the case of medicine also we have many regional traditions. One of the important markers of this is seen in the practice of the Four Glorious Tantras of medicine, and it is said that King Trisong Detsen and Acharya Padmasambhava concealed the treasures that could only be discovered later.[37] In the Nyingma tradition of Tibetan Buddhism that is closest to the teachings of Padmasambhava, there is a tradition of the treasure literature known as *terma* and a class of monks who try to recover these concealed treasures are known as *tertons* or treasure hunters.[38] This is a situation in medicine that has very few parallels in any part of the world and points to the uniqueness of the Tibetan situation.

Yuthok Yonthan Gonpo was undoubtedly one of the greatest physicians of the early period in Tibet, and there was also another famous physician of this period named Nyapa Choesang.[39] In the Tibetan Buddhist tradition, the names of individual lamas are important as they belong to a lineage, and in the case of physicians also this held true. Some of the important lineages of physicians were the Jangpa, Zurkhar and Gongmen lineages along with the lineage founded by Yuthok Yonthan Gonpo.[40] It was a lineage of Yuthok Yonthan Gonpo whose descendants in the 13th century by the same name

Yuthok Yonthan Gonpo (1126–1202) is credited with the composition of the famous text *rgyud-bzhi*.[41]

As Tibet was politically fragmented from the 9th century to the 13th century, there is no semblance of any central authority, and therefore, it was these lineages who continued the tradition of practising medicine, and thus, there was no standardization. However, there was also a degree of uniformity as medicine was part of the monastic educational curriculum, and as the literary tradition was part of the Tibetan great tradition, there was uniformity in terms of language where standard literary Tibetan was used and in the texts that were printed using the block printing. With regard to the spoken language and the folk practices, there are wide differences and variations in the regions. One of the landmark events in the political history of Tibet was unification in the 17th century under Gelugpas and the institution of Dalai Lama.

A major shift that took place in the medical history of Tibet was under the rule of the fifth Dalai Lama, Ngawang Lobsang Gyatso, also known as the Great Fifth. During his period, the central government or Ganden Phodrang was established and became very effective. In 1692, three medical centres were proposed: Sorig Dropen Ling College, Drang Song Lupsai Ling College and a school at Lawang-chok inside the Potala Palace.[42] By far, the most important landmark was the establishment of Chakpori Medical College in 1696. According to the Italian Tibetologist Tucci, it was primarily this temple that also functioned as a medical centre until it was destroyed in 1959 after the communist invasion.[43] By far, the greatest development in Tibetan medical history was the establishment of the Men Tsee Khang in 1916.[44] It was established just outside Lhasa and continues to function even today in exile in India. The name is officially translated as Tibetan Medical and Astro-sciences. One of the achievements of this institution was that not only was it the first non-monastic medical college where students from both monastic and aristocratic backgrounds gathered to study, it also was one of those few institutions that could survive in exile owing largely to the system of syllabi that are loaded to adapt to the non-monastic segments.

Like the monastic education where the young novices entered at a very early age between the ages of 7 and 11, they were tutored in the basics of the Tibetan language that was highly valued and strongly emphasized.[45] Only after a high mastery of the Tibetan language, the students were allowed to go to a higher level of study and some of them branched into subjects like medicine where the course went on for seven long years. Many aspects of medicine, ranging from the memorization of text to the identification of herbs and flowers, preparation of medicine and religious practice dedicated to the medicine Buddha, were all taught. In addition, they were also given education in aspects of astrology and astronomy that are integral parts of medicine. This also included the Tantric ritual, another magic ritual process, which was part of the healing process. An important element of medical education was the method of pulse analysis that took quite a long time to master. Many students had to spend a good number of years to complete their education, as it would be deemed complete only if they attained perfection in the pulse analysis.[46] The medicinal analysis also remained incomplete without mastery of astrology and astronomy, and therefore, the Indic sciences formed the core of the holistic knowledge system.

It was the institutional base of Tibetan monasticism and the tradition of translation by the class of translator monks that helped in the propagation and sustenance of Indic medicine in the Tibetan scenario. Thus, Ayurveda attained a new lease of life after the destruction of the great monasteries of India, and worse now, a part of a new hybrid tradition suited to the Tibetan topography.

Notes

1 Harry Ritter, *Dictionary of Concepts in History*, Greenwood Publishing Group, Boston, MA, 1986, pp. 193–200.
2 Razak Khan, "The Social Production of Space and Emotions in South Asia", *Journal of the Economic and Social History of the Orient*, Leiden, Vol. 58, No. 5, 2015, pp. 611–33 (p. 612).
3 Ruth Finnegan, *Oral Literature in Africa*, Open Book Publishers, Cambridge, UK, 2012, pp. 3–5.
4 C. Vijayalakshmi, *Making of the Past in South India: Historians Craft under Colonialism*, unpublished PhD thesis, School of Social Sciences, Mahatma Gandhi University, Kottayam, India, 2003, p. 1.
5 Ruth Finnegan, *op. cit.*, pp. 97–100.
6 Rosalind O'Hanlon, "Performance in a World of Paper: Puranic Histories and Social Communication in Early Modern India", *Past & Present*, Oxford, UK, Vol. 219, No. 1, 2013, pp. 89, 124.
7 Mathew Kapstein, "The Indian Literary Identity in Tibet", in Sheldon Pollock (ed.), *Literary Cultures in History: Reconstructions from South Asia*, University of California Press, Berkeley, CA, 2003, pp. 747–751.
8 *Ibid.*, p. 788.
9 M. N. Rajesh, *Gompas in Traditional Tibetan Society*, Decent Books, New Delhi, 2002, p. 35.
10 Lama Jabb, *Oral and Literary Continuities in Modern Tibetan Literature: The Inescapable Nation*, Lexington Books, London, 2015, pp. 10–13.
11 Theresia Hofer, *Bodies in Balance: The Art of Tibetan Medicine*, University of Washington Press, New York, 2018, p. 55.
12 David Macdonald, *Cultural Heritage of Tibet: A Description of a Country of Contrasts and of its Cheerful Happy-Go-Lucky People of Hardy Nature and Curious Customs, Their Religion, Ways of Living, Trade, and Social Life*, Light & Life Publishers, New Delhi, 1978, p. 29.
13 Antoinette K. Gordon, *Tibetan Religious Art*, Courier Corporation, New York, 2012, pp. 81, 92.
14 Anton Schiefner, *Tibetan Tales Derived from Indian Sources*, Trübner & Company, London, 1882, p. 97.
15 Bhagwan Dash, "Ayurveda in Tibet", *The Tibet Journal*, Dharamsala, Vol. 1, No.1, 1975, p. 99.
16 M. N. Rajesh, *op. cit.*, p. 51.
17 Bhagwan Dash, *op. cit.*, p. 95.

18 Sienna R. Craig and Barbara Gerke, "Naming and Forgetting Sowa Rigpa and the Territory of Asian Medical Systems", *Medicine Anthropology Theory*, Edinburgh, Vol. 3, No. 2, 2016, p. 102.
19 Thomas Laird, *The Story of Tibet: Conversations with the Dalai Lama*, Open Road + Grove Atlantic, New York, 2007, p. 133.
20 Tsering Shakya, *Whither the Tsampa Eaters? Himal Southasian*, Kathmandu, September 1, 1993, pp. 9–11.
21 Alexander Andreyev, "Russian Buddhists in Tibet, from the End of the Nineteenth Century – 1930", *Journal of the Royal Asiatic Society*, Cambridge, Vol. 11, No. 3, 2001, p. 350.
22 Graham Coleman, *A Handbook of Tibetan Culture: A Guide to Tibetan Centres and Resources Throughout the World*, Rupa and Co, New Delhi, 2016, pp. 261, 268.
23 M. N. Rajesh, *op. cit.*, p. 160.
24 Bhagwan Dash, *op. cit.*, pp. 94–104 (p. 100).
25 Christa Kletter and Monika Kriechbaum, *Tibetan Medicinal Plants*, CRC Press, London, 2001, p. 4.
26 *Ibid.*
27 Subodh Kapoor, *The Buddhists: Encyclopaedia of Buddhism*, Cosmo Publications, New Delhi, 2000, p. 1401.
28 Wang Furen, Wenqing Suo and Wen-ch'ing So, *Highlights of Tibetan History*, New World Press, Beijing, 1984, p. 18.
29 John Powers, *Introduction to Tibetan Buddhism*, Snow Lion Publications, Boulder, CO, 2007, p. 146.
30 Mathew Kapstein, *op. cit.*, pp. 760–761.
31 Tsepak Rigzin, "Rinchen Zangpo: The Great Tibetan Translator (958–1055 AD)", *The Tibet Journal*, Dharamsala, Vol. 9, No. 3, 1984, pp. 28–37 (p. 29).
32 Jampa Gyaltsen Drakton and Lobsang Chomphel, *Fundamentals of Tibetan Medicine: According to the Rgyud-Bzhi*, Tibetan Medical Centre, Dharamsala, 1981, p. 92.
33 Bhagwan Dash, *op. cit.*, p. 99.
34 John Powers and David Templeman, *Historical Dictionary of Tibet*, Scarecrow Press, Lanham, MD, 2012, pp. 602–603.
35 Theresia Hofer, *Bodies in Balance: The Art of Tibetan Medicine*, University of Washington Press, New York, 2018, p. 174.
36 Thomas Laird, *op. cit.*, p. 58.
37 Ringu Tulku, *The Ri-Me Philosophy of Jamgon Kongtrul the Great: A Study of the Buddhist Lineages of Tibet*, Shambhala Publications, Boulder, CO, 2007, p. 118.
38 Thomas Laird, *op. cit.*, p. 58.
39 Desi Sangye Gyatso, translated by Gavin Kilty, *Mirror of Beryl: A Historical Introduction to Tibetan Medicine*, Wisdom Publications, Somerville, MA, 2010, p. 37.
40 *Ibid.*, pp. 311–321.
41 *Ibid.*, p. 252.
42 Pasang Yonten, "A History of the Tibetan Medical System", *Tibetan Medicine Issue 12*, LTWA, Dharamsala, 1989, p. 46.
43 John Powers and David Templeman, *op. cit.*, p. 596.
44 Stephan Kloos, "Celebrating the Men-Tsee-Khang's 100th Anniversary: 12–13 September 2016, A Report from Lhasa", *The Tibet Journal*, Dharamsala, Vol. 43, No. 1 Spring/Summer, 2018, p. 169.
45 M. N. Rajesh, *op. cit.*, p. 51.
46 *Ibid.*

10 Tradition of Tibetan Buddhism and Neuro-psychobiology

Arun Kumar Jha

This chapter intends to present the nature of the mind and related issues as they are understood within Tibetan Buddhism as well as by neuro-psychobiologists. From a Buddhist perspective, the reason for engaging in such an investigation is simply to gain greater knowledge about the mind and learn the nature of transformation in the mind. Moreover, very little attention has so far been given to this aspect. The philosophy of *Aryastyacatusatya* (the Four Noble Truths or Tibetan: *Dhenpa Zhi*) and *Atthangikomaggo* (the Noble Eightfold Path or Tibetan: *Phaglam yanlag gye*), as enumerated in *Dhammacakkappavattanam*, is a way to transform the mind (Tibetan: *blo*) and achieve freedom from *dukkha* (suffering or Tibetan: *dhuk ngal*).[1] Such transformation can be brought about by the mind which is the area of study of the neuro-psychobiologists also; for example, they study the nature and functioning of the brain and related organs of the body. In fact, they are exploring the contact of sense stimuli and receptors, and the resulting nerve impulses sent to the brain. On the other hand, Buddhism throws light on studying the experience to which these nerve impulses and the philosophy of science of mind give rise. It believes to have a closer parallel between the science of mind and neuro-psychological activity in human beings. It is within this model that the concept of science of mind appears to be synonymous with the neuro-psychobiologists' concept of the processes that occur in one's brain.

History shows that the classical Tibetan Buddhist thought gained tremendous insights from its long association with other schools of thought in ancient India. In this regard, His Holiness, the XIV Dalai Lama, also envisions dual benefits from this engagement between classical Tibetan Buddhist thought and contemporary thought, especially neuro-psychobiology. On the one hand, he believes that a dynamic encounter with scientific thought could help revitalize Buddhist analysis of the nature of objective reality and the mind. As for the neuro-psychobiologists, this correlation with Buddhist philosophy provides new perspectives on their different topics and other disciplines. Dalai Lama further sees the potential for mutual enrichment in the scientific investigation of the chemical and neuro-physiological changes that occur in meditative states.[2] This ultimately leads us to

DOI: 10.4324/9781003370642-11

a thorough understanding of the nature and relationship between the body and mind.[3] Even *Madhyamaka* Buddhist philosophical ideas may help to deal with many of the conceptual challenges posed by neuro-biology, psychology, cosmology and quantum physics with regards to the question of reality.[4] A close perusal of the mind and life (Tibetan: *sem* and *lus*) presents an evidence of dialogues between Buddhist thought and these four fields of science.[5]

As is well known among the Four Noble Truths, the first one addresses the nature of suffering. The reason for this is because we are averse to suffering, and this subject is taught in terms of feelings. The forms of sufferings are three in nature: blatant suffering or feeling of pain, suffering of change or a tainted feeling of pleasure and ubiquitous suffering of conditioning or having neither pleasure nor pain (or indifferent). All these three types of sufferings pertain to feeling as it is directly related to the nature of consciousness, which also includes feelings and recognition.

The Second Noble Truth of the Origin of Suffering is related to mental afflictions and to *kamma* (Tibetan: *las*) or actions caused by mental agonies. These traits are considered as expressions of consciousness.[6] The *Abhidharmakosabhasya* is right when it tells us, "*Karmajam lokavaicitrayam*" or "the world in its variety arises from action".[7]

The Third Noble Truth of Cessation may also be regarded as an attribute of consciousness or realization. Within Buddhist tantric tradition or *Vajrayana*, there are classifications of different degrees of subtlety of consciousness (Tibetan: *she spa*). For example, there is a threefold classification, namely waking consciousness, dreaming consciousness and the consciousness of dreamless sleep. Its study shows the existence of a relationship between the mind and brain in relation to these various degrees of subtlety of consciousness. There is another classification which distinguishes the mind from the mental factors. The mind apprehends the sheer presence or nature of its object, whereas the mental factors apprehend specific attributes of the apprehended object. Besides, the *Sautrantika* philosophical school and all of the higher philosophical systems, namely the *Yogacara* and *Madhyamaka* schools, assert that consciousness apprehends its object by way of images. The Fourth Noble Truth of the Path to Cessation requires excellent qualities of the mind or of consciousness, specifically those qualities that lead to liberation. (Tibetan: *tharpa*). It is, therefore, called the Way or Noble Way; *Ariya-Magga* in Pali texts and *Dhenpa Zhi* in Tibetan works.[8]

Despite a great deal of debate, the most predominant thoughts of Tibetan Buddhism maintain that all phenomena are of the nature of the mind.[9] Hence, the Buddha has placed great emphasis on the mind. He is of the view that if one trains the mind (Tibetan: *lojong*), there is joy, and if the mind is indisciplined, there is suffering.[10] That, the principal things which must be purified are the contaminations of the mind, and these also are mental. And that which purifies the mind are the excellent qualities or states of mind.[11]

From a Buddhist perspective, the reason for engaging in such an investigation is not only to gain greater knowledge about the mind but also to bring about its transformation so that one may overcome the negative forces in one's emotions that create mental unrest and suffering.[12] Therefore, Buddhism known as *chos*, the Tibetan word for *Baudh dharma*, has the connotation of protecting or saving one from fears and frustrations, and sorrows and sufferings.

In fact, the depth and richness of the ways in which mental factors are put into action reach an extraordinary degree of refinement in the *Vajrayana*.[13] Its practice involves improving the quality of one's mind (Tibetan: *blo*). It means simply that one empowers those mental processes that are beneficial to oneself and disempowers those that are harmful to oneself. Besides, most of the dharmas, as enumerated in the *Abhidharma* systems, are mental. Besides, the overriding Buddhist concern with liberation from suffering, *Abhidharma* (Tibetan: *tharpa* from *dhuk ngal*), is primarily an analysis of our psychological experience. It is part of the basic philosophy of Buddhism, common to all schools – the Theravadins, the Tibetans, etc. It is also part of the *Tipitaka*. The Tipitakas are handbooks of how to live in terms of the awakened state of mind and are concerned with how to deal with actual situations. The *Abhidharma* is a very important part of those general instructions.[14]

The Buddha saw the cure for human suffering and showed the way leading to the cure of mental disorder, which is a general term for psychogenic mental disturbances. We know that under the heading of mental disorders come phobias – the neuroses and the psychoses. The common mental disorder called *dukkha* or suffering (Tibetan: *sdug sngal*) is a sort of common neurosis. Buddhism claims not only to relieve *dukkha* (the suffering) both totally and irreversibly, but also it helps to get rid of the development of any phobia, neurosis or psychosis for life. When the suffering is cured or the mind is free from the clutches of delusion (Tibetan: *marig-pa*), the resulting state of experience is *nibban* or liberation (Tibetan: *tharpa*), which denotes the state of sound mental health[15] and the awakened state of mind, and thus to be free from the world (Tibetan: *khorba*) of life and death. It is not a theoretical state, existing only as an ideal or an unrealized possibility. Thousands of people have actually attained *nibban* (Tibetan: *tharpa*) in this way.[16] It appears also to be the same, wherever it occurs, and not contaminated by any local cultural bias.

Thus, suffering is the disorder and liberation the state of order. The Buddhist therapy to concern with the alleviation and cure of suffering and the attainment of emancipation is achieved by following the *lam rim*, a Tibetan word, which means to follow a graded path to wisdom and compassion (Tibetan: *sherab* and *jampa*) through which one can train one's mind (Tibetan: *lojong*) in order to get enlightenment.[17]

The whole therapy of mental transformation is also embodied in the Noble Eightfold Path (Tibetan: *Phaglam yanlag gye*). It can be divided, on the one hand, into preparatory or auxiliary therapy, and on the other, the therapy proper. The preparatory therapy includes a regime of everyday behaviour and mental discipline and an introduction to Buddhist theory and philosophy. The therapy proper is *dhyan* or meditation (Tibetan: *sgom*).[18]

The preparatory therapy, on the basis of its characteristics, may be categorized into three major divisions[19] and is represented in the form of a chart given in Figure 10.1:

This Eightfold Path, applied to conditions of contemporary life, is a self-therapy for a happy and healthy life. It does not require us to renounce the world (Tibetan: *khorba*). It only demands modifying our approach to life (Tibetan: *lus*), and through rigorous mental discipline, we may move on the path to self-realization and thus transform the mind (Tibetan: *sem*).

In the Buddhist works, the *Satipatthana* and the *Mahasatipatthana Suttas* besides the *Milinda Panha*, it has been said that the human body is a composite of various organs consisting of liquid and solid materials. From the neuro-biological and biochemical points of view, this body is nothing but a

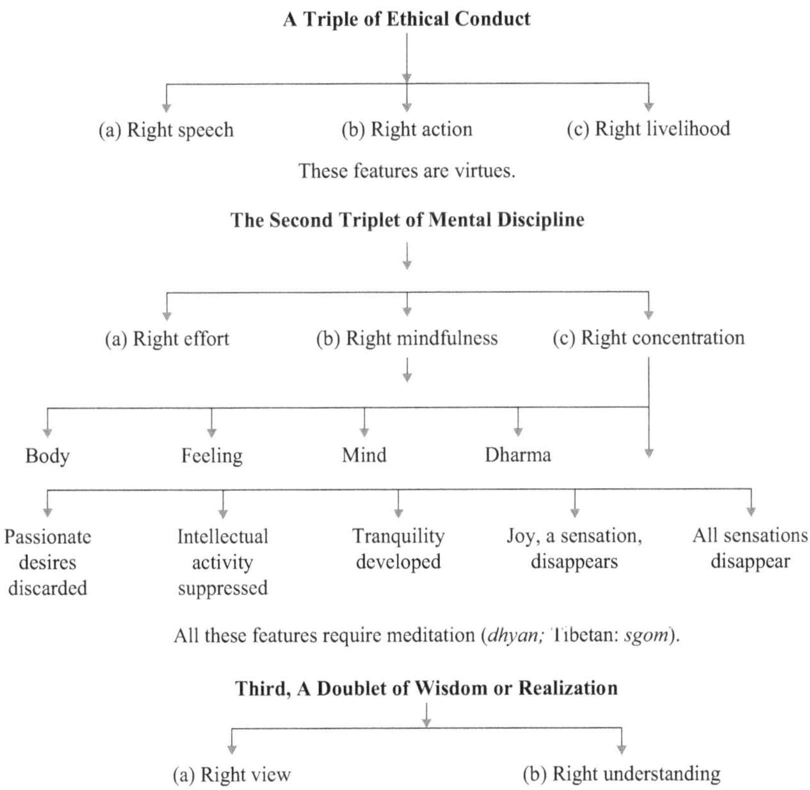

Figure 10.1 The Characteristics of Preparatory Therapy

composition of the organic system in which chemical processes are undergoing constant transformation.

To keep the body and mind in a harmonious state is, according to Buddhist teachings, the prerequisite for the maintenance of physical, emotional and mental health. This harmonious balance is brought about by the practice of virtue, the First Triplet of Ethical Conduct. Virtue is described in the *Visuddhimagga* as abstention, volition and non-transgression. Impediments to virtue are anger, jealousy, pride, arrogance, lust and non-practice of restraint of the senses. The states that counteract these are the causes of virtue. It is important to note that any of these states produce alterations in the neuro-biological balance of the body. The more unequilibrated one becomes, the more neurotransmitter substances such as Acetylcholine, Serotonin and Epinephrine are released. These trigger the production of many different types of effects on body organs such as the thyroid, adrenal cortex, digestive tracts and reproductive organs. For example, when rage, anger or agitation is experienced, the production of Epinephrine causes the blood pressure to rise and cardiac output and oxygen consumption to increase.[20] Other common states that we experience are fear and anxiety. During these states, far more than the normal amount of acid required is excreted into the intestinal tracts, whereby the membrane is attacked resulting in ulcers. Prolonged states of nervous tension are known to impair the autonomic nervous system in such a way that it influences negatively peristalsis, which impairs digestive functions and elimination.[21] This results in constipation and self-poisoning, thus adversely affecting the physical, emotional and mental health of the individual. Mental health is defined "as a mind freed from the influence of the afflictive mental factors", and that is the goal of the process of meditation.[22]

It can be seen why Buddha stressed the practice of virtue as a prerequisite for meditation. The practice of meditation is impossible when the physical body, emotions and mind of the individual are impaired. One who has practised virtue can successfully pursue meditation and discard the hindrances which would finally pave the way for sound health and spiritual upliftment or liberation. Some of the vital aspects of meditation, the therapy proper, remain unique to Tibetan Buddhism for transforming the mind. We come to know that meditation has neuro-psychobiological effects on long-term meditators – it can help in preventing the development of neuroses and psychoses.

The Tibetan word for meditation is *sgom*. It means to familiarize with a positive state of mind, which actually refers to training the mind (Tibetan: *lojong*).[23] It is *gtong len*, or the practice of giving and taking, which differentiates Tibetan Buddhist meditation most clearly from the Indian form of meditation. '*Tongi*' means 'sending out' or 'letting go', and '*len*' means 'taking in'.[24] So, the 'practice of giving and taking' (Tibetan: *gtong len*) is the basis of this *bodhicitta* (Tibetan: *byangs chub sems*) field training. This is the supreme practice to gain compassion and have virtuous acts which lead to enlightenment. In this practice, one should think that one's own mother sits in front of oneself. Then one should recount all the good things one's mothers showered on the self. Then one needs to think that all sentient beings have been our mothers and have showered us with love, care and support just as our present mother has. Now one has to think that one should give them the same in return.[25]

Since all sentient beings suffer from some sort of neuroses and psychoses which lead to stress, anxiety and depression, one should develop the *bodhicitta* mind to accept one's suffering as a container and give them happiness in return. One should visualize and accept the suffering of all sentient beings in the form of black smoke through their nostrils. Whereas, when one starts to exhale, one should visualize wholeheartedly that one's entire happiness and prosperity is being emitted through one's nostrils to the hearts of suffering sentient beings. This practice of giving and taking in Tibetan is known as *gtong len*.

The philosophy underlying Tibetan *gtong len* identifies the very nature of mind, its innermost essence shadowed by jumbled thoughts and afflictive emotions (Tibetan: *yon-mongs*). This can be overcome by meditation. Proper meditation of the *zhi-gnas* (Sanskrit: *shamatha*) or 'tranquillity meditation' and *lhag-mthong* (Sanskrit: *vipashyana*) or 'insight meditation' can lead to perfectly realized human beings. In fact, when we work only for our self-interest, our potential cannot be fully developed because we are very limited by such a narrow view of life. Instead, if we open up our mind to see and care for the well-being of all the sentient beings, our ability will gradually grow and our potential will develop to full maturity. The origin of our spiritual impurity is our narrow-minded selfishness.[26]

In this way, the mind training (Tibetan: *lojong*) results in the development of compassion and love of the Buddha, for ourselves and others as well.[27] It helps to change our neuro-biosystem and enhance the physical and spiritual awareness through which we learn to be mindful, which is the most important basic human need. From this perspective of Buddhism, the faculties of the mind (Tibetan: *sem*) facilitate the task of its transformation through the pursuit of knowledge (Tibetan: *shesja*), meditation (Tibetan: *sgom*), love (Tibetan: *nying jey*) and compassion (Tibetan: *jampa*).

Notes

1 B. J. Kashyap (ed.), *The Mahavagga*, PPB, Bihar Government, Patna, 1956, p. 13.

2 The inaugural lecture was presented for the Fourteenth Dalai Lama Endowment for Tibetan Buddhism and Cultural Studies at the University of California in the USA. on 2 June 1997; also see Mr. Tenzin Geyche, the Private Office of His Holiness the Dalai Lama, Mcleodganj, HP.

3 Dalai Lama et al., *Mind Science: An East-West Dialogue,* (eds.) D. Goleman and R. A. F. Thurman, Paperback Edition, Wisdom Publications Inc., Boston, MA, 1991, Ch. I, Ch. III.

4 See, for example, Dalai Lama, *The Four Noble Truths*, Thorsons Publishing Group Ltd., London, 1998, Ch. VI–VII.
5 For further explanations, see Dalai Lama and Thubten Chodron, *Approaching the Buddhist Path*, Wisdom Publications Inc., MA, USA, 2020, pp. xi–xiii, 3, 17, 61, 63.
6 The Buddhists and neuro-biologists largely concur that the very forms and structures of human life result from the accumulative actions of innumerable beings over countless generations. D. Snellgrove, *Indo-Tibetan Buddhism: Indian Buddhists and Their Tibetan Successors*, Shambhala Publications Inc., Boston, MA, 2002, pp. 1–4, 18–35.
7 S. D. Shastri (ed.), *Adbhidharmakosa-bhasya*, Vol. 1, Baudha Bharati, Varanasi, 1998, *dukham samudaya loko drishtisthanam*, VAKK-1.8, 1.15, 1.16, 1–8c – VAKK-1.811.
8 Carol S. Anderson, *Pain and Its Ending: The Four Noble Truths in the Theravada Buddhist Cannon*, under Curzon critical studies in Buddhism series, Routledge, 2013, Ch. I, Ch. V.
9 This is the *Prasaigika* School.
10 Dalai Lama, "Understanding and Transforming the Mind", in B. Alan Wallace (ed.), *Buddhism & Science*, Motilal Banarsidass (Henceforth MLBD), Delhi, 2004, p. 93.
11 B. D. Kyokai, *The Teaching of Buddha*, Kosaido Printing Co. Ltd., Tokyo, Japan, 1979, pp. 15, 46–74.
12 Walpola Rahula, *What the Buddha Taught*, MLBD, Delhi, 2017, Ch. 1, Ch. VIII.
13 F. J. Varela and N. Depraz, "Imagining Embodiment, Phenomenology and Transformation", in B. Alan Wallace (ed.), *op. cit.*, p. 219.
14 Carol S. Anderson, *op. cit.*, Ch. IV.
15 T. T. Drungtso, *Healing Power of Mantra,* Drungtso Publications, Dharmasala, 2006, p. 138.
16 G. Du Pre, "Buddhism and Psychotherapy", in B. P. Kirthisinghe (ed.), *Buddhism & Science*, MLBD, Delhi, 1999, p. 98.
17 T. T. Drungtso, *op. cit.,* p. 139.
18 H. R. Beech, *Changing Man's Behaviour*, Pelican, 1969, pp. 23–29.
19 Shanti Tayal, "The Buddha's Middle Way and Western Psychotherapy", in B. P. Kirtisinghe, *op. cit.,* p. 43.
20 U. Aung Thien, "Buddhist Meditation and Bioscience", in B. P. Kirthisinghe (ed.), *op. cit.*, p. 57.
21 *Ibid.*
22 Walpola Sri Rahula, *History of Buddhism in Ceylon*, Atlantic, New York, 1959, Ch. I, V, VII, VIII.
23 T. T. Drungtso, *op. cit.,* p. 84.
24 F. J. Varela and N. Depraz, *op. cit.*, in B. Alan Wallace (ed.), *op. cit.*, p. 221.
25 T. T. Drungtso, *op. cit.*, p. 140.
26 Arun Kumar Jha, "Historiography on Tibetan Science of Buddhist Medicine", in *Proceedings of Indian History Congress*, 74th Session, Cuttack, 2013 (Delhi, 2014), pp. 719–722.
27 Sogyal Rinpoche, *The Tibetan Book of Living and Dying*, Harper Collins, London, 1992, Ch. IV–V. This book is the author's best-selling work in recent years.

Select Bibliography

Primary Sources

Vedic Scriptures

Griffith, Ralph T. H. (trs.). 1971. *The Hymns of the Rgveda.* 2 Vols. Varanasi: The Chowkhamba Sanskrit Series Office (Reprint).
Sharma, S. R. and B. D. Sharma (ed.). 1983. *Rg-Veda Samhita.* 4 Vols. New Delhi: Motilal Banarsidass Publishers Pvt. Ltd.
———. 1985. *Atharvaveda-Samhita.* 2 Vols. New Delhi: Motilal Banarsidass Publishers Pvt. Ltd.
Whitney, William Dwight (trs.) and Charles Rockwell Lanman (ed.). 1971. *Atharva-Veda-Samhita.* 2 Parts. New Delhi: Motilal Banarsidass Publishers Pvt. Ltd.

Brahmanas and Upanishads

Eggling, Julius (trs.). 1963. *The Satapatha–Brahamana According to the Text of the Madhyandina School.* 5 Vols. Delhi: Motilal Banarsidass.
Keith, A. B. (trs.). 1971. *The Aitareya and Kausitaki Brahmanas of the Rg-Veda.* New Delhi: Motilal Banarsidass (Reprint).
Mitra, Rajendralal (ed.). 1859–1962. *The Taittiriya Brahmana of the Black Yajur Veda with the Commentary of Sayanacarya.* 3 Vols. Calcutta: Asiatic Society of Bengal.
Muller, F. Max (trs.). 1962. *The Upanishads.* 2 Vols. New York: Dover Publications.

Smritis and Sutras

Dvivedi, Kesawa Prasad Sarma. 1867. *Manu-Smriti.* Bombay: Laxmi Venkteswar Press.
Gharpure, J. R. (trs.). 1936. *Yajnavalkya-Smriti.* Bombay: Published by V. J. Gharpure, at the office of the Collection of Hindu Law Texts, Girgaon.
Jolly, Julius (ed.). 1962. *Vishnu Smriti*, Varanasi: Chowkhamba.

Puranas

Pargiter, F. E. (trs.). 1969. *Markendya Purana.* Varanasi: Chowkhamba.

Wilson, H. H. (trs.) and Fitzedward Hall (ed.). 1877. *The Vishnu Purana: A System of Hindu Mythology and Tradition*. 6 Vols. London: Trubner & Co.

Buddhist Texts

Kashyap, B. J. (ed.). 1956. *The Mahavagga*. Patna: PPB, Bihar Government.

Shastri, S. D. (ed.). 1981. *Adbhidharmakosabhasyam*. Varanasi: Chowkhamba.

Other Texts and Translations

Abhyankara, Kasinath Vasudeva (ed.). 1975. *Mahabhashya of Patanjali*. Poona: Bhandarkar Oriental Research Institute.

Ayurvedacraya, Sri Taradattapanta (ed.). 1956. *Astangahrdayasamhita of Vagbhatta*. Calcutta: The Chowkhamba Sanskrit Series Office.

Bhisagratna, K. K. (ed.). 1963. *The Susruta Samhita* (English translation based on the original Sanskrit texts). Varanasi: Chowkhamba Sanskrit Series Office.

Burgess, Revd. E. 1935. *Translation of the Surya Siddhanta*. Edited by P. Gangooly and P. Sengupta. Calcutta: Calcutta University.

Chatterjee, Bina (ed.). 1970. *The Khandakhādyaka (An Astronomical Treatise) of Brahmagupta: Introduction, Translation and Mathematical Notes*. Vol. 1. Calcutta: World Press.

———. 1981. *Sisyadhivriddhida Tantra of Lalla with Commentary of Mallikarjuna Suri: Part 2: Translation and Mathematical Notes*. New Delhi: Indian National Science Academy.

Dutt, Uday Chand (trs.). 1883–91. *The Susruta-Samhita: The Hindu System of Medicine According to Susruta*. Calcutta: Asiatic Society of Bengal.

Ray, Priyadaranjan and Hirendra Nath Gupta. 1965. *Caraka Samhita: A Scientific Synopsis*. New Delhi: The National Institute of Sciences of India.

Sarup, Lakshman (trs.). 1967. *The Nighantu and The Nirukta*. New Delhi: Motilal Banarsidass.

Sastri, V. S. Venkatasubramania and C. Raja Rejisevara Sarma (eds.). 1977. *The Bhela Samhita*. New Delhi: Central Council for Research in Indian Medicine and Homeopathy.

Sastry, T. S. Kuppanna and K. V. Sarma (ed.). 1985. *Vedanga Jyotisa of Lagadha in its Rk and Yajus Recensions: With the Translation and Notes of Prof. T. S. Kuppanna Sastry*. New Delhi: Indian National Science Academy.

Sharma, Ram Karan and Vaidya Bhawan Dash (trs.). 1976, 1977. *Agnivesa's Carak Samhita*. 1&2. Calcutta: The Chowkhamba Sanskrit Series Office.

Shukla, Kripa Shankar (ed.). 1959. *Patiganita of Sridharacharya*. Lucknow: Lucknow University.

———. 1976. *Aryabhatiya of Aryabhata*. Part I. New Delhi: Indian National Science Academy (INSA).

———. 1976. *Aryabhatiya of Aryabhata*. Part III. New Delhi: Indian National Science Academy (INSA).

Manuscripts

Bakshali Manuscript. Bodleian Library. London: University of Oxford.

Hoernle, A. F. Rudolf (ed. & trs.). 1893–1897. *The Bower Manuscripts*. 7 Parts. Calcutta: Office of the Superintendent of Government Printing, India (ASI).

Secondary Sources

Abbas, R., M. Vahia, V. Shobha, R. K. Mohanty and A. Sule. 2010. *Search for Archaeoastronomical Data from the Megalithic Burials of Junapani, Nagpur, Maharashtra*. Bombay: Tata Institute of Fundamental Research.

Agrawal, D. P. 2009. *Harappan Technology and its Legacy*. New Delhi: Rupa.

Allchin, Bridget and Raymond Allchin. 1968. *The Birth of Indian Civilization*. Harmendsworth: Penguin Books Ltd.

———. 1982. *The Rise of Civilization in India and Pakistan*. Cambridge: Cambridge World Archaeology, Cambridge University Press.

Amin, Samir. 1989. *Eurocentrism*. New York: Russell Moore and James Membrez, NYU Press.

Anderson, Carol S. 2015. *Pain and its Ending: The Four Noble Truths in the Theravada Buddhist Cannon*. Under Curzon critical studies in Buddhism series. London: Routledge.

Andrews, Munya. 2004. *The Seven Sisters of the Pleiades: Stories from Around the World*. North Melbourne, VIC: Spinifex Press.

Ashfaque, Syed Mohammad. 1992. *Primitive Astronomy in the Indus Civilization in Old Problems and New Perspectives in the Archaeology of South Asia*. Madison, WI: Department of Anthropology, University of Wisconsin.

Avari, Burjor. 2016. *India: The Ancient Past: A History of the Indian Subcontinent from c. 7000 BCE to 1200 CE*. New York: Routledge, Second edition.

Aveni, Anthony F. 1981. "Archaeoastronomy". In Michael B. Schiffer (ed.). *Advances in Archaeological Method and Theory*. Vol. 4. Cambridge, MA: Academic Press, Inc.

———. 1989. *Empires of Time: Calendars, Clocks, and Cultures*. New York: Basic Books Inc.

Bag, A. K. (ed.). 1997. *History of Science & Technology*. Vol. I. New Delhi: Indian National Science Academy (INSA).

Baishya, Dinesh. 2009. *Traditional Science and Material Culture of Early Assam*. Guwahati: EBH Publishers.

Banarjee, N. R. 1965. *The Iron Age in India*. Delhi: Munshiram Manoharlal.

Bandyopadhyay, Arun (ed.). 2010. *Science and Society in India (1750–2000)*. New Delhi: Manohar.

Basa, K. K. and P. Mohanty (eds.). 2000. *Archaeology of Orissa*. New Delhi: Pratibha Prakashan.

Basham, A. L. 1976. "The Practice of Medicine in Ancient and Medieval India". In Charles Leslie (ed.), *Asian Medical Systems*. Berkeley, CA: University of California Press.

———. 2004. *The Wonder that was India*. London: Picador.

Beech, H. R. 1969. *Changing Man's Behaviour*. Baltimore, MD: Pelican, Penguin Books.

Benice, Daniel D. 1978. *Mathematics, Ideas and Applications*. New York: Academic Press.

Bhardwaj, H. C. 1979. *Aspects of Ancient Indian Technology*. Delhi: Motilal Banarsidass.

Biswas, A. K. and Sulekha Biswas. 1996. *Minerals and Metals in Ancient India*. New Delhi: D. K. Printworld.
Bose, D. M., et al. (eds.). 1971. *A Concise History of Science in India*. New Delhi: Indian National Science Academy (INSA).
Brunn, Ole and Arnae Kalland (eds.). 1955. *Asian Perceptions of Nature*. London: Curzon Press.
Carey, Bertram S. and H. N. Tuck. 2008. *The Chin Hills*. Vol. I & II. Calcutta: Firma KLM.
Chakrabarti, Dilip K. 1993. *The Early Use of Iron in India*. Delhi: Oxford University Press (OUP).
———. 1999. *INDIA: An Archaeological History*. New Delhi: Oxford University Press (OUP).
———. 2006. *The Oxford Companion to Indian Archaeology*. New Delhi: Oxford University Press (OUP).
Chattopadhyaya, D. P. 1979. *Science and Society in Ancient India*. Calcutta: Research India Publication.
——— (ed.). 1982. *Studies in the History of Science in India*. Vol. II. New Delhi: Editorial Enterprises.
———. 1986. *History of Science and Technology in Ancient India: The Beginnings*. Calcutta: Firma KLM Pvt. Ltd.
Chattopadhyaya, D. P. and Ravinder Kumar (eds.). 1995. *Science, Philosophy and Culture in Historical Perspective*. New Delhi: PHISPC.
———. 1995. *Some Aspects of India's Philosophical and Scientific Heritage*. New Delhi: PHISPC.
Chattopadhyay, Pranab K. and Gautam Sengupta. 2011. *History of Metals in Eastern India and Bangladesh*. New Delhi: Pentagon Press.
Chaudhary, V. C. P. 1977. *Secularism versus Communalism*. Patna: Navdhara Samiti.
Clark, Walter Eugene (trs.). 1930. *Aryabhata: An Ancient Indian Work on Mathematics and Astronomy*. Chicago, IL: University of Chicago Press.
Clifford, T. 1984. *Tibetan Buddhist Medicine and Psychiatry - The Diamond Healing*. Maine (USA): Samuel Weiser, Inc.
Colebrooke, H. T. 1817. *Algebra with Arithmetic and Mensuration from the Sanscrit of Brahmagupta and Bhascara*. London: Cambridge University Press.
Coleman, Graham. 2016. *A Handbook of Tibetan Culture: A Guide to Tibetan Centres and Resources throughout the World*. New Delhi: Rupa and Co.
Courant, Richard and Robbins Herbert. 1969. *What is Mathematics?* New York: Oxford University Press, Inc.
Craufurd, Quintin. 1790. *Sketches Chiefly Relating to the History, Religion, Learning and Manners of the Hindoos with a Concise Account of the Present State of the Native Powers of Hindostan*. London: T. Cadell.
Crombie, A. C. (ed.). 1963. *Scientific Change*. London: Heinemann.
Danda, Ajit K. 1996. *Tribal Ethnography*. New Delhi: ICSSR.
Dani, S. G. 1993. *Vedic Mathematics: Myth and Reality*. New Delhi: Economic and Political Weekly.
Darrah, Z. H., et al. 1896. *Notes on Some Industries of Assam*. Shillong: Assam Secretariat Printing Office.
Das, Vaidya Bhagwan. 2001. *Materia Medica of Ayurveda*, Based on Madanapala's *Nighantu*. New Delhi: B. Jain Publishers Pvt., Ltd.

Dasgupta, Nupur and Amit Bhattacharyya (eds.). 2014. *Essays in History of Science, Technology and Medicine*. Kolkata: Setu Prakashani.

Dasgupta, S. N. 1922. *History of Indian Philosophy*. Vol. III. Cambridge University Press.

Datta, Bibhutibhushan. 1932. *The Science of the Sulba*. Calcutta: University of Calcutta.

Dhavalikar, M. K. 2002. *Environment and Culture: A Historical Perspective*. Pune: Bhandarkar Oriental Research Institute.

Drakton, Jampa Gyaltsen and Lobsang Chomphel. 1981. *Fundamentals of Tibetan Medicine: According to the Rgyud-Bzhi*. Dharamsala: Tibetan Medical Centre.

Drungtso, T. T. 2006. *Healing Power of Mantra*. Dharamsala: Drungtso Publications.

Duke-Elder, W. S. (ed.). 1969. *Systems of Ophthalmology*. St Louis, MO: CV Mosby.

Dummer, Tom. 2001. *Tibetan Medicine and Other Holistic Healthcare Systems*. New Delhi: Paljor Publications.

Elgood, C. 1951. *A Medical History of Persia*. Cambridge: Cambridge University Press.

Elwin, Verrier. 1942. *The Agaria*. London: Oxford University Press (OUP); Humphrey Milford.

———. 1993. *Myths of the North-East Frontier of India*. Itanagar: Directorate of Research Govt. of Arunachal Pradesh.

Evans, James. 1998. *The History & Practice of Ancient Astronomy*. New York: Oxford University Press (OUP).

Filliozat, Jean. 1964. *The Classical Doctrine of Indian Medicine: Its Origins and its Greek Parallels*. Translated by Devraj Channana. Delhi: Munshiram Manoharlal.

Finnegan, Ruth. 2012. *Oral Literature in Africa*. Cambridge: Open Book Publishers.

Francis, John. 2008. *Philosophy of Mathematics*. New Delhi: Global Vision Publishing House.

Furen, Wang, Wenqing Suo, and Wen-ch'ing So. 1984. *Highlights of Tibetan History*. Beijing: New World Press.

Gait, E. A. 1905. *A History of Assam*. London: Thacker Spink & Co.

Garnayak, D. B. 2012. *Early Historic Archaeology of Odisha*. Bhubaneswar: Mayur Publications.

Garrett, Frances. 2008. *Religion, Medicine and the Human Embryo in Tibet*. Routledge.

Gopalakrishnan, N. 2000. *Indian Scientific Heritage*. Trivandrum: Indian Institute of Scientific Heritage.

Gordon, Antoinette K. 2012. *Tibetan Religious Art*. New York: Courier Corporation.

Gordon, B. L. 1949. *Medicine throughout Antiquity*. Philadelphia, PA: A. Davis Co.

Goswami, Priyam. 2005. *Indigenous Industries of Assam: Retrospect and Prospect*. Published for Maulana Abul Kalam Azad Institute of Asian Studies, Kolkata. Delhi: Anshah Publishing House.

Gurdon, P. R. T. 1906. *The Khasis*. London and New Delhi: Akansha Publishing House (Reprint 2010).

Gurukkal, Rajan and Raghava M. R. Varier. 1999. *Cultural History of Kerala*. Thiruvananthapuram: Dept. of Cultural Publications, Govt. of Kerala.

Gyatso, Desi Sangye (translated by Gavin Kilty). 2010. *Mirror of Beryl: A Historical Introduction to Tibetan Medicine*. Sommerville, MA: Wisdom Publications.

Gyllenbok, Jan. 2018. *Encyclopaedia of Historical Metrology, Weights, and Measures*. Vol. 1. *Science Networks Historical Studies* 56. Cham: Springer International Publishing (Birkhauser).

Habib, Irfan. 2008. *Medieval India: The Study of a Civilization*. New Delhi: National Book Trust.

Habib, S. Irfan and Dhruv Raina (eds.). 1999. *Situating the History of Science: Dialogue with Joseph Needham*. New Delhi: Oxford University Press (OUP).

———. 2007. *Social History of Science in Colonial India*. New Delhi: Oxford University Press (OUP).

Hamilton, Buchanan. 1963. *An Account of Assam*. Guwahati: Department of Historical and Antiquarian Studies.

Hayward, J. W. and F. J. Varela (eds.). 1992. "*Gentle Bridges: Conversations with the Dalai Lama on the Science of Mind*". In *Mind and Life*. Virginia: Shambhala Publications.

Heggie, D. C. (ed.). 1982. *Archaeoastronomy in the Old World*. Cambridge: Cambridge University Press.

Hobsbawm, Eric. 2011. *How to Change the World: Tales of Marx and Marxism*. London: Little Brown.

Hoernle, A. F. R. 1907. *Studies in the Medicine of Ancient India Part I: Osteology or the Bones of the Human body*. Oxford: Clarendon Press, Oxford Press.

Hofer, Theresia. 2018. *Bodies in Balance: The Art of Tibetan Medicine*. New York: University of Washington Press.

Hogben, Lancelot. 1965. *Mathematics for the Million*. New York: Pocket Books.

Hunter, W. W. 1879. *A Statistical Account of Assam*. London: Trubner & Co.

Hutton, J. H. 1995. *Diaries of Two Tours in the Un-administered Area East of the Naga Hills*. New Delhi: Mittal Publications.

Jabb, Lama. 2015. *Oral and Literary Continuities in Modern Tibetan Literature: The Inescapable Nation*. London: Lexington Books.

Jaggi, O. P. 1969. *Dawn of Indian Technology*. Delhi: Atma Ram & Sons.

———. 1981. *Indian System of Medicine*. Delhi: Atma Ram & Sons.

Joseph, George Gheverghese. 2016. *Indian Mathematics: Engaging with the World from Ancient to Modern Times*. London: World Scientific Publishing Europe Ltd.

Joseph, Tony. 2018. *Early Indians*. New Delhi: Juggernaut Books.

Kakoty, Sanjeeb. 2003. *Technology Production and Social Formation in the Evolution of the Ahom State*. New Delhi: Regency.

Kanigel, Robert. 1991. *The Man who Knew Infinity*. New York: Washington Square Press Pocket Books.

Kapoor, Subodh. 2000. *The Buddhists: Encyclopaedia of Buddhism*. New Delhi: Cosmo Publications.

Keith, Arthur Berriedale. 1928. *A History of Sanskrit Literature*. Walton Street: Clarendon Press.

Kelley, David H. and Eugene F. Milone. 2011. *Exploring Ancient Skies: A Survey of Ancient and Cultural Astronomy*. New York: Springer, Second edition.

Kennedy, Kenneth A. R. 2000. *God-Apes and Fossil Men*. Ann Arbor, MI: University of Michigan Press.

———. 2001. "Middle and Late Pleistocene Hominids of South Asia". In Phillip V. Tobias, et al. (eds.). *Humanity from African Naissance to Coming Millennia: Colloquia in Human Biology and Palaeoanthropology*. Firenze: Firenze University Press; Johannesburg, ZA: Witwatersrand University Press.

Keshwani, N. H. (ed.). 1974. *The Science of Medicine and Physiological Conceptions in Ancient and Medieval India*. New Delhi: Dept. Of Physiology. All-India Institute of Medical Sciences.

Kirthisinghe, B. P. (ed.). 1999. *Buddhism & Science*. Delhi: Motilal Banarsidass.

Kletter, Christa and Monika Kriechbaum. 2001. *Tibetan Medicinal Plants*. London: CRC Press.

Kosambi, D. D. 1965. *The Culture and Civilization of Ancient India in Historical Outline*. London: Routledge.

Kragh, Helge. 1987. *Introduction to Historiography of Science*. Cambridge: Cambridge University Press.

Kumar, Deepak. 1995. *Science and Raj: 1857–1905*. New Delhi: Oxford University Press (OUP).

Kuppuram, G. 1989. *Ancient Indian Mining, Metallurgy and Metal Industries*. Delhi: Sundeep Prakashan.

Kuppuram, G. and K. Kumudamini (eds.). 1990. *History of Science and Technology in India*. 12 Vols. Delhi: Sudeep Prakashan.

Kutumbiah, P. 1962. *Ancient Indian Medicine*. New Delhi: Orient Longmans.

Kyokai, B. D. 1979. *The Teaching of Buddha*. Tokyo: Kosaido Printing Co. Ltd.

Laird, Thomas. 2007. *The Story of Tibet: Conversations with the Dalai Lama*. New York: Grove/Atlantic.

Lama, Dalai and Thubten Chodron. 2017. *Approaching the Buddhist Path*. Somerville. MA: Wisdom Publications Inc.

Lama, Dalai, et al. 1991. *Mind Science: An East-West Dialogue*. D. Goleman and R. A. F. Thurman (eds.). Boston, MA: Paperback Edition, Wisdom Publications Inc.

Lama, Dalai. 1998. *The Four Noble Truths*. London: Thorsons Publishing Group Ltd.

Lankford, John (ed.). 2011. *History of Astronomy: An Encyclopaedia*. New York: Routledge.

Leach, E. R. 1970. *Political Systems of Highland Burma: A Study of Kachin Social Structure*. London: University of London, The Athlone Press.

Lehman, F. K. 1980. *The Structure of Chin Society*. Aizawl-Mizoram: Firma KLM Private Ltd on behalf of Tribal Research Institute.

Leuva, K. K. 1963. *The Asur: A Study of the Primitive Iron Smelters*. New Delhi: Bharatiya-Adimjati-Sevak Samaj.

Lindsay, L. 1849. *Lives of Lindsay or a Memoir of the House of Crawford and Balcarres*. Vol. III. London: John Murray.

Macdonald, David. 1978. *Cultural Heritage of Tibet: A Description of a Country of Contrasts and of its Cheerful Happy-Go-Lucky People of Hardy Nature and Curious Customs, their Religion, Ways of Living, Trade, and Social Life*. New Delhi: Light & Life Publishers.

Mackenzie, Alexander. 1884. *History of the Relations of Government with the Hill Tribes of the North-East Frontier of Bengal*. Calcutta: The Home Department Press.

Mackie, Thomas T., et al. 1945. *A Manual of Tropical Medicine*. Philadelphia, PA: W. B. Suanders Company.

Magli, Giulio. 2016. *Archaeoastronomy*. Undergraduate Lecture Notes in Physics. Cham: Springer International Publishing.

Majumdar, G. P. 1951. *The History of Botany and Allied Sciences in Ancient India*. Paris: Archives Internationales d' Historire des Sciences, UNESCO.

Mallik, Ashok Kumar. 2018. *The Story of Number*. Singapore: IISc Press-World Scientific.
Malsawmliana and Benjamin Ralte (eds.). 2011. *Social, Economic and Political History of the Mizo*. Guwahati: EBH Pub. (India).
Marshack, Alexander. 1991. *The Roots of Civilization: The Cognitive Beginnings of Man's First Art, Symbol and Notation*. New York: Moyer Bell.
McDowell, F. 1977. *The Source Book of Plastic Surgery*. Baltimore, MD: William & Wilkins Co.
Menon, C. P. S. 1932. *Early Astronomy and Cosmology: A Reconstruction of the Earliest Cosmic System*. London: George Allen and Unwin Ltd.
Mirsky, Jeannette. 1977. *Sir Aural Stein: Archaeological Explorer*. Chicago, IL: University of Chicago Press.
Moorti, Udayaravi S. 1994. *Megalithic Culture of South India: Socioeconomic Perspectives*. Varanasi: Ganga Kaveri Pub. House.
Mukhopadhyay, Dhruba. 2016. *Science in Ancient India - Reality vs Myth*. Kolkata: Breakthrough Science Society.
Mukhopadhyaya, Girindranath. 1913. *The Surgical Instruments of the Hindus*. 2 Vols. Calcutta: Calcutta University Press.
———. 1914. *History of Indian Medicine*. 3 Vols. Calcutta: Orient Books.
———. 1974. *History of Indian Medicine*. Vol. I. New Delhi: Oriental Books Reprint Corporation.
———. 1994. *Ancient Hindu Surgery: Surgical Instruments of the Hindus*. New Delhi: Cosmo Publications (Reprint).
Nariman, G. K., M. Winternitz, S. Levi, and E. Huber. 1972. *Literary History of Sanskrit Buddhism*. New Delhi: Motilal Banarsidass (Reprint).
Narrien, John. 1833. *Historical Account of the Origin and Progress of Astronomy*. London: Baldwin Cradcock.
Nehru, Jawaharlal. 2004. *The Discovery of India*. New Delhi: Penguin Books.
Neuberger, M. 1910. *History of Medicine*. Translated by E. Playfair. Vol. I. London: Oxford University Press (OUP).
Nilsson, Martin P. 1920. *Primitive Time-Reckoning; A Study in the Origins and First Development of the Art of Counting Time among the Primitive and Early Culture Peoples*. Lund and Oxford: Oxford University Press.
O' Connor, J. J. and E. F. Robertson. 2000. *Aryabhata the Elder*. Scotland: University of St.Andrews.
Pannekoek, A. 1961. *A History of Astronomy*. London: George Allen &Unwin.
Parpola, Asko. 1988. *Religion Reflected in the Iconic Signs of the Indus Script: Penetrating Into Long-Forgotten Picto+graphic Messages*. Leiden, Holland: Brill.
———. 1994. *Deciphering the Indus Script*. New York: Cambridge University Press.
———. 2010. *A Dravidian Solution to the Indus Script Problem*. Chennai: Central Institute of Classical Tamil.
Patniak, Prabhat (ed.). 2011. *Excursus in History: Essays on Some Ideas of Irfan Habib*. New Delhi: Tulika Books.
Petraglia, Michael D. and Bridget Allchin (eds.). 2007. *The Evolution and History of Human Populations in South Asia: Inter-Disciplinary Studies in Archaeology, Biological Anthropology Linguistics and Genetics*. Netherlands: Springer.
Pingree, David. 1994. *A Census of the Exact Sciences in Sanskrit*. 5 Vols. Philadelphia, PA: American Philosophical Society.
Plofker, Kim. 2009. *Mathematics in India*. Princeton, NJ: Princeton University Press.

Pollock, Sheldon (ed.). 2003. *Literary Cultures in History: Reconstructions from South Asia*. Berkeley, CA: University of California Press.

Powers, John and David Templeman. 2012. *Historical Dictionary of Tibet*. Lanham, MD: Scarecrow Press.

Powers, John. 2007. *Introduction to Tibetan Buddhism*. Boulder, CO: Snow Lion Publications.

Pradhan, S. (ed.). 2006. *Art and Archaeology of Orissa: Recent Perspectives*. New Delhi: Aryan International.

Prakash, U. 1978. *Shushruta of Ancient India*. Sur Gyn and Obs. Chicago, IL: The Franklin H. Martin Memorial Foundation.

Rahman, Abdur, et al. 1982. *The Monograph Science and Technology in Medieval India: A Bibliography of Source Materials in Sanskrit, Arabic and Persian*. New Delhi: Indian National Science Academy.

Rahman, Abdur (ed.). 1984. *Science and Technology in Indian Culture: A Historical Perspective*. New Delhi: NISTADS.

Rahula, Walpola Sri. 1959. *History of Buddhism in Ceylon*. New York: Atlantic.

———. 2017. *What the Buddha Taught*. Delhi: Motilal Banarsidass.

Raina, Dhruv. 2003. *Images and Contexts: The Historiography of Science and Modernity of India*. New Delhi: Oxford University Press (OUP).

Rajesh, M. N. 2002. *Gompas in Traditional Tibetan Society*. New Delhi: Decent Books.

Rao, S. Balachandra. 2000. *Indian Astronomy: An Introduction*. Hyderabad: Universities Press.

———. 2004. *Indian Mathematics and Astronomy: Some Landmarks*. Bangalore: Bhavan's Gandhi Centre of Science and Human Values.

Ray, P. and Hirendra Nath Gupta. 1965. *Caraka Samhita: A Scientific Synopsis*. New Delhi: The National Institute of Sciences of India.

Ray, P. C. (ed.). 1956. *History of Chemistry in Ancient and Medieval India*. Calcutta: Asiatic Society.

———. 2002. *A History of Hindu Chemistry*. Centenary Edition. Kolkata: Shaibya Prakashan Bibhag.

Renfrew, Colin and Iain Morley (eds.). 2010. *The Archaeology of Measurement: Comprehending Heaven, Earth and Time in Ancient Societies*. Cambridge: Cambridge University Press.

Rinpoche, Sogyal. 1992. *The Tibetan Book of Living and Dying*. London: Harper Collins.

Robinson, William. Reprint. 1975. *A Descriptive Account of Assam*. New Delhi: Sanskaran Prakashak.

Roller, H. D. (ed.). 1971. *Perspectives in the History of Science and Technology*. Oklahoma: Norma.

Rosen, Frederic August. 1831. *The Algebra of Mohammed Ben Musa*. Edited and translated by Frederic Rosen. London: Oriental Translation Fund.

Roy, A. and Bagchi, S. K. (eds.). 1986. *Technology in Ancient and Medieval India*. New Delhi: Sandeep Prakashan.

Roy, Keka Dutta and Chitta Ranjan Mishra (eds.). 2009. *Reflections in History: Essays in Honour of Professor Amalendu De*. Kolkata: Raktakarabee Publisher.

Royle, J. F. 1837. *An Essay on the Antiquity of Hindoo Medicine*. London: William H. Allen & Co.

Ruggles, Clive L. N. (ed.). 2015. *Handbook of Archaeoastronomy and Ethnoastronomy*. New York: Springer.

Sahu, B. P. (ed.). 2006. *Iron and Social Change in Early India*. New Delhi: Oxford University Press (OUP).
Sankalia, Hasmukhlal Dhirajlal. 1962. *The Prehistory and Protohistory of India and Pakistan*. Bombay: University of Bombay.
Sarkar, B. K. 1937. *India in Exact Science: Old and New*. Calcutta: Indian Research Institute.
Sarkar, Susobhan. 1970. *Mysticism in the History of Science: Bengal Renaissance and Other Essays*. New Delhi: People's Publishing House.
Sarton, G. A. L. 1975. *Introduction to the History of Science*. Vol. I. From Homer to Omar Khayyam. Huntington, NY: Robert E. Krieger Publishing Company.
Schiefner, Anton. 1882. *Tibetan Tales Derived from Indian Sources*. London: Trubner & Company.
Scott, J. F. 1967. *A History of Mathematics*. London: Taylor & Francis.
Seal, B. 1915. *The Positive Sciences of the Ancient Hindus*. London: Longmans, Green and Co.
Selin, Helaine and Rodam Narasimha (eds.). 2007. *Encyclopaedia of Classical Indian Sciences: Natural Science, Technology, Medicine*. Hyderabad: University Press.
Sen, Amartya. 2005. *The Argumentative Indian*. London: Penguin Books.
Sen, S. N. and K. S. Shukla. 1985. *History of Astronomy in India*. New Delhi: Indian National Science Academy (INSA).
Sengupta, G., S. Roychoudhury, and S. Som (eds.). 2006. *Past and Present Ethnoarchaeology in India*. Delhi: Pragati Publications.
Settar, S. and Ravi Korisettar (eds.). 2002. *Indian Archaeology in Retrospect: Prehistory - The Early Archaeology of South Asia*. Vol. 1. New Delhi: ICHR & Manohar.
Shadap-Sen, N. C. 1981. *The Origin and Early History of the Khasi-Synteng People*. Calcutta: Firma KLM Pvt. Ltd.
Sharma, P. V. (ed.). 1992. *History of Medicine in India*. New Delhi: Indian National Science Academy (INSA).
Sharma, R. S. 2005. *India's Ancient Past*. New Delhi: Oxford University Press (OUP).
Siddiqi, M. Z. 1959. *Studies in Persian and Arabic Medical Literature*. Calcutta: Calcutta University.
Singh, Bal Ram, et al. (eds.). 2012. *Science and Technology in Ancient Indian Texts*. New Delhi: D. K. Printworld (P) Ltd.
Singh, J. P. and G. Sengupta (eds.). 1991. *Archaeology of North Eastern India*. New Delhi: Har Anand.
Singh, Upinder and Nayanjyot Lahiri (eds.). 2009. *Ancient India: New Research*. New Delhi: Oxford University Press (OUP).
Singhal, G. D., S. N. Tripathy, and G. N. Chaturvedi (trs.). 1981. *Ancient Indian Surgery*. Varanasi: Banaras Hindu University.
Smellgrove, D. 2002. *Indo-Tibetan Buddhism: Indian Buddhists and Their Tibetan Successors*. Boston, MA: Shambhala Publications Inc.
Snow, C. P. 1959. *The Two Cultures and a Second Look*. London: Cambridge University Press.
Srivathsa, B. K. and M. A. Narasimhan (ed.). 2003. *Science and Technology in India through the Ages*. Melkote: Academy of Sanskrit Research.
Staal, Frits. 1993. *Concepts of Science in Europe and Asia*. Leiden: IIAS.

Stevenson, H. N. C. 1986. *The Economics of the Central Chin Tribes*. Aizawl: Tribal Research Institute, Directorate of Education, Govt. of Mizoram.

Subbarayappa, B. V. (ed.). 1997. *Chemistry and Chemical Techniques in India*. New Delhi: Oxford University Press (OUP).

Subbarayappa, B. V. and K. Venkateswara Sarma. 1985. *Indian Astronomy: A Source-Book Based Primarily on Sanskrit Texts*. Bombay: Nehru Centre.

Sundara, A. 1975. *The Early Chamber Tombs of South India: A Study of the Iron Age Megalithic Monuments of North Karnataka*. Delhi: University Publishers (India).

Syiemlieh, David R. 2000. *Survey of Research in History of North-East India (1970–1990)*. New Delhi: Regency.

Temple, Sir Richard C. 1903. *The Andaman and Nicobar Islands: Report on the Census, Census of India 1901*. Vol. III. Calcutta: Office of the Superintendent of Government Printing.

Thakur, Vijay Kumar and K. K. Mandal (eds.). 2000. *Science, Technology and Medicine in Indian History: Essays in Honour of Dr.Devendra Kumar Singh*. Patna: Janaki Prakashan.

Thapar, Romila. 1965. *A History of India*. Vol. I. London: Penguin Books.

———. 2004. *Cultural Pasts: Essays in Early Indian History*. New Delhi: Oxford India Paperbacks.

Thorwald, J. 1963. *Science and Secrets of Early Medicine*. Translated by R. Winston and C. Winston. New York: Harcourt, Brace & World Inc.

Thurston, Hugh (ed.). 1996. "The Astronomy of Aryabhata" in *Early Astronomy*. New York: Springer.

———. 1996. *Early Astronomy*. New York: Springer.

Tripathy, B. and D. V. Prasad (eds.). 2016. *Ethnoscience and Traditional Technology in India*. New Delhi: Aayu Publication.

Tripathy, Vibha. 2008. *History of Iron Technology in India (From Beginning to Pre-modern Times)*. New Delhi: Rupa & Co.

Trungpa, Chogyam. 2001. *Glimpses of Abhidharma*. Boston, MA: Paperback Shambhala Dragon Edition.

Tulku, Ringu. 2007. *The Ri-Me Philosophy of Jamgon Kongtrul the Great: A Study of the Buddhist Lineages of Tibet*. Boulder, CO: Shambhala Publications.

Walker, Christopher (ed.). 1996. *Astronomy before the Telescope*. London: British Museum Press.

Wallace, B. Alan (ed.). 2004. *Buddhism & Science*. Delhi: Motilal Banarsidass.

Whipple, A. O. 1963. *The Story of Wound Healing Repair*. Springfield, IL: Charles C.Thomas.

Winter, Henry James Jacques. 1952. *Eastern Science: An Outline of its Scope and Contribution*. London: John Murray (Reprint).

Wise, T. A. 1860. *Commentary on the Hindu System of Medicine*. London: Trubner & Co.

Wujastyk, Dominik. 2003. *The Roots of Ayurveda: Selections from Sanskrit Medical Writings*. New Delhi: Penguin Books.

———. 2009. *Mathematics and Medicine in Sanskrit*. Delhi: Motilal Banarsidass.

Yadav, B. S. and Man Mohan (eds.). 2010. *Ancient Indian Leaps into Mathematics*. New York: Springer.

Yonten, Pasang. 1989. *A History of the Tibetan Medical System Tibetan Medicine*. Issue 12. Dharamsala: LTWA.

Zimmer, H. R. 1948. *Hindu Medicine*. H. N. Lectures. Baltimore, MD: John Hopkins Press.
Zysk, K. G. 1993. *Religious Medicine: The History and Evolution of Indian Medicine*. Routledge.
———. 1998. *Medicine in the Veda: Religious Healing in the Veda*. Delhi: Motilal Banarsidass.
———. 2000. *Asceticism and Healing in Ancient India: Medicine in the Buddhist Monastery*. Delhi: Motilal Banarsidass.

Journals and Reports

Ancient India, ASI, New Delhi
Ancient India, New Delhi
Annals of Bhandarkar Oriental Research Institute, Pune
Annals of Medical History, USA
Asiatic Researchers, Calcutta
Bulletin of the Astronomical Society of India, Hyderabad
Centaurus, United Kingdom
Chopra Committee Reports, 1948
Current Science, Bangalore
Dave Committee Reports, 1955
Economic and Political Weekly, New Delhi
Ethnohistory, USA
Heritage: Journal of Multidisciplinary Studies in Archaeology, Kerala
Historia Mathematica, Netherland
History of Science in South Asia (HSSA Journal), Canada
Indian Journal of History of Medicine
Indian Journal of History of Science, New Delhi
Indian Journal of Ophthalmology
Indian Journal of Plastic Surgery
Indian Journal of Surgery
International Institute for Asian Studies Newsletter, Netherland
International Journal of Humanities and Social Science, USA
Journal for the History of Astronomy, UK
Journal of Astronomical History and Heritage, Thailand
Journal of Astronomy and Astrophysics, Egypt
Journal of Ayurveda
Journal of Indian Archaeological Society, New Delhi
Journal of Royal Asiatic Society Bulletin, Bombay
Journal of Social Archaeology, Sage Journals
Journal of the American Oriental Society, USA
Journal of the Economic and Social History of the Orient, Leiden
Journal of the Pakistan Historical Society, Karachi
Journal of the Royal Asiatic Society of Bengal, Calcutta
Journal of the Royal Asiatic Society, Cambridge
Journal of the Royal Society of Arts
Journal of Tropical Medicine & Hygiene, UK

Man and Environment, Pune
Man in India, New Delhi
Man, UK
Nature, UK
Orissa Historical Research Journal, Bhubaneswar
Past & Present, Oxford, UK
Proceedings of Indian History Congress
Proceedings of International Conference on Traditional Asian Medicine, Canberra, 1979
Proceedings of the 2nd International Conference on Tibetan Medicine
Proceedings of the Seventh International Conference on Oriental Astronomy
Report of the meeting of the Panel on Ayurveda, Planning Commission, Govt. of India, 1960
Science, USA
The Indian Historical Review, ICHR, New Delhi
Udupa Committee Reports, 1958
Vistas in Astronomy

Dictionary and Encyclopaedia

Dictionary of Concepts in History, Boston
Dictionary of Scientific Biography, New York
Encyclopaedia of the History of Science, Technology and Medicine in Non-Western Cultures, Springer, Netherland

Index

For Product Safety Concerns and Information please contact our EU
representative GPSR@taylorandfrancis.com
Taylor & Francis Verlag GmbH, Kaufingerstraße 24, 80331 München, Germany

www.ingramcontent.com/pod-product-compliance
Ingram Content Group UK Ltd.
Pitfield, Milton Keynes, MK11 3LW, UK
UKHW020138200926
475702UK00005B/803

* 9 7 8 1 0 3 2 2 4 8 5 9 2 *